Praise for
Fifty Things that Made the Modern Economy **and**
The Next Fifty Things that Made the Modern Economy

'Endlessly insightful and full of surprises – exactly what you would expect from Tim Harford'

Bill Bryson

'Short chapters are a delight in this frenetic age ... Best of all, the book is constantly surprising. It brims with innovations I didn't know about, as well as ones I thought I knew about but did not'

The Times

'Packed with fascinating detail ... Harford has an engagingly wry style and his book is a superb introduction to some of the most vital products of human ingenuity'

Sunday Times

'Splendid ... Harford is a fine, perceptive writer, and an effortless explainer of tricky concepts. His book teems with good things, and will expand the mind of anyone lucky enough to read it'

Daily Mail

'Harford's richness of detail bespeaks skill both as an economic analyst and as a popular commentator'

Times Literary Supplement

'Tim Harford, always excellent, turns his eye towards inventions. The plough, the gramophone, the pill, the Billy bookcase. Over and over, Harford shows us, inventions have all sorts of knock-on effects'

Evening Standard

D1375301

By Tim Harford

THE NEXT
FIFTY THINGS
THAT MADE THE
MODERN ECONOMY

Tim Harford

The
Bridge
Street
Press

THE BRIDGE STREET PRESS

First published in Great Britain in 2020 by The Bridge Street Press
This paperback edition published in 2021 by The Bridge Street Press

1 3 5 7 9 10 8 6 4 2

Copyright © Tim Harford 2020

The moral right of the author has been asserted.

All rights reserved.
No part of this publication may be reproduced, stored in a
retrieval system, or transmitted, in any form or by any means, without
the prior permission in writing of the publisher, nor be otherwise circulated
in any form of binding or cover other than that in which it is published
and without a similar condition including this condition being
imposed on the subsequent purchaser.

A CIP catalogue record for this book
is available from the British Library.

ISBN 978-0-3491-4403-0

Typeset in Bembo by M Rules
Printed and bound in Great Britain by
Clays Ltd, Elcograf S.p.A.

Papers used by The Bridge Street Press are from well-managed forests
and other responsible sources.

The Bridge Street Press
An imprint of
Little, Brown Book Group
Carmelite House
50 Victoria Embankment
London EC4Y 0DZ

An Hachette UK Company
www.hachette.co.uk

www.littlebrown.co.uk

To Fran

Contents

IV: Invisible Systems

V: Secrets and Lies

VI: Working Together

VII: No Planet B

VIII: Our Robot Overlords

INTRODUCTION

1

The Pencil

When Henry David Thoreau, the great nineteenth-century American essayist, made a comprehensive list of supplies for an excursion, he specified obvious items such as a tent and matches, added string, old newspapers, a tape measure and a magnifying glass, and also included paper and stamps, to make notes and write letters. Strange, then, that he omitted to mention the very pencil with which he was making the list.[1] Stranger still, when you realise that Thoreau and his father made their money by manufacturing high-quality pencils.[2]

The pencil seems fated to be overlooked, which makes it the ideal subject of an old English riddle: 'I am taken from a mine, and shut up in a wooden case, from which I am never released, and yet I am used by almost everybody.' Nobody declares that 'the pencil is mightier than the sword' – not so long as erasers exist.

But overlooked is just the way I like it. I am an admirer of the things that tend to pass unnoticed. From the brick to the 'Like' button, cellophane to the menstrual pad, the inventions that fill the pages of this book are often taken for granted.

Their stories are seldom told, and the lessons they might hold are rarely learned. That is why – I hope – those stories and lessons might teach us more than a discussion of more obvious breakthroughs such as the steam engine or the computer.

In selecting the fifty-one subjects of this book, my aim has been to tell stories that will surprise you, about ideas that have had fascinating consequences. There are plenty of other books about inventions that changed the world; this book is about inventions that might change the way you see that world.

And there is no better place to start than with the poor neglected pencil. We don't even give it the courtesy of a sensible name. 'Pencil' is derived from the Latin word *penis*, which means – yes, yes, settle down – 'tail'. That is because Roman writing brushes were made from tufts of fur from an animal's tail.

'Lead pencils' achieve the same effect without needing ink. Or, indeed, lead – because pencil leads are actually made of graphite. The idea of graphite on a stick of wood is about 450 years old. Yet more than two centuries later, *Encyclopaedia Britannica* was still defining a 'pencil' as a brush, just as Cicero or Seneca might have used.[3]

But the pencil does have some champions. Henry Petroski, a historian of the pencil, points out that its very erasability makes it indispensable to designers and engineers. In his words, 'Ink is the cosmetic that ideas will wear when they go out in public. Graphite is their dirty truth.'[4]

And then there's the American economist Leonard Read, who was a crusader for the principles of small-government free-market economics. In 1958, Read published an essay titled 'I, Pencil' – written in the voice of the pencil itself. While the pencil in the English riddle sounded resigned to its obscurity, Read's pencil is a proselytising libertarian with a melodramatic disposition: 'if you can become aware of the miraculousness which I symbolize, you can help save the freedom mankind is so unhappily losing'.[5]

Read's pencil is well aware that it doesn't at first appear impressive: 'Pick me up and look me over. What do you see? Not much meets the eye—there's some wood, lacquer, the printed labeling, graphite lead, a bit of metal, and an eraser.'

And yet, the pencil goes on to explain, collecting its cedar wood required saws, axes, motors, rope and a railway car; its graphite comes from Ceylon – modern-day Sri Lanka – mixed with Mississippi clay, sulphuric acid, animal fats and numerous other ingredients. And don't get the pencil started on its six coats of lacquer, its brass ferrule, or its eraser – made not from rubber, it wants you to know, but from sulphur chloride reacted with rape-seed oil, made abrasive with Italian pumice and tinted pink with cadmium sulphide.[6]

And what is the ingenious answer to the perennial question, how do they get the graphite inside the wood? The trick is to take a slim slab of that cedar wood, kiln dried, and saw a row of grooves into the top surface. Originally the grooves were square – easier to cut by hand. Now they are precision-machined with a semicircular cross-section.[7] Once the cylindrical rods of graphite are laid into the grooves, glue another grooved slab on top – this time with the grooves in the bottom – and then cut the whole graphite sandwich into sticks, parallel to the graphite rods. These sticks are unformed pencils, so plane, varnish, and the job is done.[8]

All this to produce a pencil that can be yours for pennies – a box of 150 retails for £14.99 – and to which most of us give not a moment's thought.

But Read's plucky pencil is undaunted. It draws a stirring conclusion from the complexity of its international supply chains and the precision of its manufacture:

'Leave all creative energies uninhibited ... Have faith that free men and women will respond to the Invisible Hand. This faith will be confirmed.'[9]

Read's essay became famous when the economist Milton Friedman – a Nobel memorial prize winner, free-market champion and a gifted communicator of economic ideas – adapted it for his 1980 TV series *Free to Choose*. Friedman drew the same lesson from the humble pencil's formidably complex origins; it was an astonishing testimony to the power of market forces to coordinate large numbers of people with nobody in overall charge:

'There was no commissar sending out orders from central office; it was the magic of the price system.'[10]

Go back in time five hundred years or so, and you'd have seen the magic of the price system swing into action. Graphite was first discovered in the English Lake District. Legend has it that a ferocious storm uprooted trees in the idyllic valley of Borrowdale. Underneath their roots was a strange, shiny black substance that was initially dubbed 'black lead'.[11] Did it have any uses that would justify investing in a mine? Well, yes. Graphite was promptly used as 'a marking stone', as celebrated in this London street hawker's cry from three centuries ago:

> Buy marking stones, marking stones buy
> Much profit in their use doth lie;
> I've marking stones of colour red,
> Passing good, or else black Lead.[12]

Because graphite was soft yet heat-resistant it was also used for casting cannonballs. It soon became a precious resource. Not quite as pricey as its mineral cousin, diamond – both are forms of carbon – but valuable enough for miners to be supervised by armed guards as they changed out of their clothes at the end of the shift, lest they try to smuggle a nugget away.[13]

By the late 1700s, French pencil manufacturers were happily paying to import high-quality Borrowdale graphite. But then war broke out, and England's government sensibly decided not to make it easy for the French to cast cannon-balls. What were the French pencil-makers to do? In stepped Nicholas-Jacques Conté, French army officer, balloonist, adventurer – and pencil engineer. Conté painstakingly developed a way to make pencil leads from a mix of clay with low-grade powdered continental graphite. For these efforts, the French government awarded him a patent.

The back story of Leonard Read's heroic pencil, it turns out, is even more complex than the pencil itself acknowledges. But some details of this back story invite us to question whether Read's pencil is right to be so fiercely proud of its free-market ancestry. Would Monsieur Conté have put such effort into his experiments without the prospect of a state-backed patent? Perhaps; perhaps not. The economist John Quiggin raises a different objection: while Read's pencil underlines its history of forests and railway carts, both forests and railways are often owned and managed by governments.[14]

And while Friedman was right that there is no Pencil Tsar, even in a free-market economy there are hierarchies. That is an insight explored by another Nobel laureate, Friedman's colleague Ronald Coase. Leonard Read's loquacious instrument was made by the Eberhard Faber company, now part of Newell Rubbermaid – and, as in any conglomerate, its employees respond to instructions from the boss, not to prices in the market.

In practice, then, the pencil is the product of a messy economic system in which the government plays a role and corporate hierarchies insulate many workers from Friedman's 'magic of the price system'. Leonard Read might well be right that a pure free market would be better, but his pencil doesn't prove it.

It does, though, remind us how profoundly complex are the processes that produce the everyday objects whose value we often overlook. The economy that assembles them for us cheaply and reliably is an astonishing thing. It defies understanding. Still, for a starting point we could do worse than examine an everyday item such as a credit card, a McDonald's hamburger, a can of baked beans, or an RFID chip on a T-shirt. With every item begins a story of unexpected connections and intriguing consequences.

In short, if we want to try to understand our modern economy, Read's pencil points the way.

I

DECEPTIVELY SIMPLE

2

Bricks

'I found Rome a city of bricks and left it a city of marble.' That is supposed to have been the boast of Caesar Augustus, the first Roman emperor, just over two thousand years ago. If it was, he was exaggerating:[1] ancient Rome is a city of brick, and no less glorious for that.

Augustus was also joining a long tradition of denigrating or overlooking one of the most ancient and versatile of building materials. The great Roman architectural writer Vitruvius mentions them only in passing.[2] Denis Diderot's great French Encyclopaedia 'of the Sciences, Arts and Crafts', published in 1751 and an inspiration for Adam Smith's famous description of the pin factory[3] – well, Diderot doesn't trouble himself to include any images of brickmaking at all.[4]

That's because a brick is such an intuitive thing: people have been teaching themselves to build simple structures out of brick for many thousands of years – and grand ones too. The Hanging Gardens of Babylon were made of brick. So was the biblical tower of Babel: 'Come, let's make bricks and bake them thoroughly.' That's Genesis 11 verse 3. 'They used brick instead of stone.'[5] By verse 5, The Lord is on the scene, unimpressed by the hubris of it all, and things aren't looking too good for the brick-loving citizens of Babel.

As James Campbell and Will Pryce point out in their magisterial history of bricks, the humble cuboid is everywhere.[6] The biggest man-made structure on the planet, the Ming Dynasty Great Wall of China, is largely constructed of brick. The astonishing temples of Bagan in Myanmar; mighty Malbork Castle in Poland; Siena's Palazzo and Florence's Duomo; the bridges of Isfahan in Iran; Hampton Court Palace in West London. All brick. So is the best church in the world, Hagia Sophia in Istanbul, and the best skyscraper, the Chrysler Building in Manhattan, and even the Taj Mahal. Brick. Brick. Brick. The architect Frank Lloyd Wright once boasted that he could make a brick worth its weight in gold.[7]

This all started a long time ago; bricks seem to have been with us since the very dawn of civilisation – the oldest were found in Jericho, in Jordan, by the archaeologist Kathleen Kenyon in 1952. They are something between 10,300 and 9600 years old, and are simply loaves of mud, baked dry in the sun, then stacked up and glued together with more mud.[8]

The next big step forward was the simple brick mould, also originating from Mesopotamia, at least 7000 years old, and depicted with great clarity on a tomb painting in Thebes, Egypt. The brick mould is a wooden rectangle, with four sides but no top or bottom, into which clay and straw could be packed to make bricks faster and more precisely. These moulds can't have been easy to make – they pre-date the use of metal itself – but once constructed they made mud bricks much cheaper and better.[9]

Even in a dry climate, sun-dried mud bricks do not usually last. Fired bricks are much more durable – they're stronger, and waterproof. Making such bricks, by heating clay and sand at a temperature of about 1000° centigrade, has been possible for many thousands of years – but at a price. Accounts from the Third Dynasty of Ur, dating back just over 4000 years, note that you could get 14,400 mud bricks for the price of a piece of silver; but only 504 fired clay bricks – an exchange rate of nearly 29 mud bricks for a single clay one. Some 1500 years later, by Babylonian times, kiln technologies had improved and the price of fired clay bricks had fallen to between 2 and 5 mud bricks.[10]

That's still too much for many people – cheap and easy mud bricks are still perhaps the most popular material in the world for building houses.[11] But, as the Nobel memorial prize-winning economists Abhijit Banerjee and Esther Duflo observe, fired bricks can be an effective way for a very poor household to save. If you have a little money, buy a brick or two. Slowly, slowly, slowly, you'll have a better house.[12]

The brick is one of those old technologies, like the wheel or paper, that seem to be basically unimprovable. 'The shapes and sizes of bricks do not differ greatly wherever they are made,' writes Edward Dobson in the fourteenth edition of his *Rudimentary Treatise on the Manufacture of Bricks and Tiles*.[13] There's a simple reason for the size: it has to fit in a human hand. As for the shape, building is much more straightforward if the width is half the length.

That's why, if you get your nose up close to some buildings that seem vibrantly distinctive to their culture – the Minaret of Kalan Mosque in Uzbekistan, Herstmonceux Castle in England, the Twin Pagodas of Suzhou in China – you'll find the bricks are all much the same. It's precisely the uniformity of the brick that makes it so versatile – a lesson freshly redis-covered by every generation of parents when their children start playing with Lego.

Lego, by the way, point out that their plastic bricks don't need to be sent for recycling because they can be reused almost indefinitely. And what is true for toy bricks is truer yet for the real thing. Lego's interlocking bricks demand a high level of precision – the fault rate is just 18 per million.[14] But bricks jointed with mortar have a higher tolerance. Many medieval buildings, such as St Albans Cathedral in England, simply reused Roman bricks. Why not?

'Bricks manage time beautifully.' That's Stewart Brand in his book and TV series, *How Buildings Learn*. 'They can last nearly forever. Their rough surface takes a handsome patina that keeps improving for centuries.'[15] My own house, a brick building from the mid-nineteenth century, now has a large glass door in the back. To make the hole for the glass, we took away some bricks. Then we mixed them with similar reclaimed bricks, and used the brick salad to extend the house elsewhere.

Brick production still uses traditional methods in many parts of the world – for example in India, handmade bricks are often fired using a Bull's Trench kiln – a long ditch lined with bricks that can burn almost any fuel and produce 30,000 bricks a day. It may be fuel-hungry and polluting, but it uses local labour and materials.[16]

But automation is gradually nosing its way into most parts of brick production: hydraulic shovels dig the clay; slow conveyor belts carry bricks through long tunnel kilns; forklift trucks shift precision-stacked pallets of bricks. All this makes the brick itself cheaper.[17]

Building sites have resisted automation: the weather and the unique demands of each site require well-trained workers. The bricklayer has long been celebrated as a symbol of the honest dignity of skilled manual labour, and bricklaying tools have barely changed since the seventeenth century. But, as in so many other professions, there are signs that the robots may be coming to bricklaying. A human bricklayer can lay

300–600 bricks a day; the designers of SAM, the Semi-Automated Mason, claim it can do 3000.[18]

What of the brick itself? Various designs of interlocking brick, much like Lego, are catching on across the developing world: the result tends to be less strong and waterproof than bricks and mortar, but they're quicker and cheaper to lay.[19] And if you have robot bricklayers, why not give them bigger hands so you can make bigger bricks? Hadrian X is a robot arm which lays gigantic bricks that no human bricklayer could wield.

Maybe we shouldn't get too excited, though. SAM has a predecessor – the 'Motor Mason', for which similar claims were made back in 1967.[20] Perhaps the bricklayer will last a little longer yet. The brick certainly will.

3

The Factory

Piedmont, in north-west Italy, is celebrated for its fine wine. But when a young Englishman, John Lombe, travelled there in the early eighteenth century, he wasn't going to savour a glass of Barolo. His purpose was industrial espionage. Lombe wished to figure out how the Piedmontese spun strong yarn from silkworm silk. Divulging such secrets was illegal, so Lombe sneaked into a workshop after dark, sketching the spinning machines by candlelight. In 1717, he took those sketches to Derby in the heart of England.[1]

Local legend has it that the Italians took a terrible revenge on Lombe, sending a woman to assassinate him. Whatever the truth of that, he died suddenly at the age of 29, just a few years after his Italian adventure.

While Lombe may have copied Italian secrets, the way he and his older half-brother Thomas used them was completely original. The Lombes were textile dealers, and seeing a shortage of the strong silk yarn called organzine, they decided to go big.

In the centre of Derby, beside the fast-flowing river Derwent, the Lombe brothers built a structure that was to be imitated around the world: a long, slim, five-storey building with plain brick walls cut by a grid of windows. It housed three dozen large machines powered by a 7-metre-high waterwheel. It was a dramatic change in scale, says the historian Joshua Freeman. The age of the large factory had begun with a thunderclap.[2]

It's a testament to the no-nonsense functionality of the Derby silk mill that it operated for 169 years, pausing only on Sundays, and for droughts – when the Derwent flowed slow and low. Over that period, the world economy grew more than fivefold,[3] and factories were a major part of that growth.

Intellectuals took note. Daniel Defoe, author of *Robinson Crusoe*, came to gaze in wonder at the silk mill. Adam Smith's *The Wealth of Nations*, published in 1776, begins with a description of a pin factory.[4] Three decades later, William Blake had penned his line about 'dark Satanic Mills'.[5]

Concerns about the conditions in factories have persisted ever since. The 'Round Mill', built in 1811 not far from the Derby silk mill, was modelled after Jeremy Bentham's famous 'Panopticon' prison, a place where you never knew whether you were being watched. The circular design did not catch on, but the relentless scrutiny of workers did.[6]

Critics claimed that factory exploitation was a similar evil to slavery – a shocking claim then and now. After visiting the mills of Manchester in 1832, the novelist Frances Trollope wrote that factory conditions were 'incomparably more severe' than those suffered by plantation slaves.[7] Indeed, the factory recruiting wagons that toured the rural areas of 1850s Massachusetts, hoping to persuade 'rosy-cheeked maidens' to come to the city to work in the mills, were dubbed the 'slavers'.[8]

Friedrich Engels, whose father owned a Manchester factory, wrote powerfully about the harsh conditions, inspiring his friend Karl Marx.[9] But Marx, in turn, saw hope in the fact that so many workers were concentrated together in one place: they could organise unions, political parties, and even revolutions. He was right about the unions and the political parties, but not about the revolutions: those came not in industrialised societies but agrarian ones.

The Russian revolutionaries weren't slow to embrace the factory. In 1913, Lenin had skewered the stopwatch-driven, micromanaging studies of Frederick Winslow Taylor[10] as 'advances in the extortion of sweat'. After the revolution, the stopwatch was in the other hand. Lenin announced: 'We must organize in Russia the study and teaching of the Taylor system.'[11]

In developed economies, the dark satanic mills gradually gave way to cleaner, more advanced factories.[12] It is the working conditions of factories in developing countries that now attract attention. Economists have tended to believe both that sweatshop conditions beat the alternative of even more extreme poverty in rural areas – and that they have certainly been enough to draw workers to fast-growing cities. Manufacturing has long been viewed as the engine of rapid economic development.[13]

So what lies next for the factory? History offers several lessons.

Factories are getting bigger. The eighteenth-century Derby Silk Mill employed three hundred workers, a radical step at a time when even machine-based labour could take place at home or in a small workshop. The nineteenth-century Manchester factories that had horrified Engels could employ more than a thousand, often women and children.[14] Modern factories in advanced economies are much larger still: Volkswagen's main factory in Wolfsburg, Germany employs over 60,000 workers; that's half the population of the town itself.[15]

And the Longhua Science and Technology Park in

Shenzhen, China – better known as 'Foxconn City' – employs at least 230,000 workers, and by some estimates 450,000, to make Apple's iPhones and many other products.[16] These are staggering numbers for a single site: the entire McDonald's franchise worldwide employs fewer than 2 million.[17]

The increase in scale isn't the only way in which Foxconn City continues the arc of history. There are – as there were in the 1830s – fears for the welfare of workers. In Shenzhen, they are dissuaded from suicide by nets designed to catch anyone who leaps from the factory roof.[18]

But Leslie Chiang, who has interviewed many Chinese factory workers, notes that they know what they're doing and don't need the guilt of Western consumers. One of them, Lu Qingmin, had developed a career in the factories, met her husband, brought up a family – and saved enough to buy a second-hand Buick. 'A person should have some ambition while she is young,' she declared.[19]

Large strikes are commonplace in China, as Marx might have predicted.[20] The Chinese government, in one of history's great ironies, is cracking down on the young Marxists who try to get the workers unionised.[21]

And as in the West many decades before, there is progress: the journalist James Fallows, who has visited 200 Chinese factories, notes that conditions have dramatically improved over time.[22]

Trade secrets kick-started the first factory, and have shaped factories ever since. Richard Arkwright, whose cotton mill was modelled on the Lombe brothers' silk mill, vowed, 'I am Determind for the feuter [future] to Let no persons in to Look at the wor[k]s.'[23] Chinese factories are still secretive: Fallows was surprised to be allowed into the Foxconn plant, but he was told that he must neither show nor mention the brand names coming off the production lines.[24]

There is one clear break from the past. Factories used to centralise the production process: raw materials came in, finished products went out. Components would be made on site or by suppliers close at hand. Charles Babbage, factory enthusiast and Victorian designer of proto-computers, pointed out that this saved on the trouble of transporting heavy or fragile objects in the middle of the manufacturing process.[25]

But today's production processes are themselves global. Production can be coordinated and monitored without the need for physical proximity, while shipping containers and bar codes streamline the logistics. Modern factories – even behemoths like Foxconn City – are just steps in a distributed production chain. Components move backwards and forwards across borders in different states of assembly.[26]

Foxconn City, for example, doesn't make iPhones: they assemble them, using glass and electronics from Japan, Korea, Taiwan and even the USA.[27] Huge factories have long supplied the world. Now the world itself has become the factory.

4

The Postage Stamp

'It should be remembered, that in few departments have important reforms been effected by those trained up in practical familiarity with their details. The men to detect blemishes and defects are among those who have not, by long familiarity, been made insensible to them.'[1]

Those words are from 1837. An early pitch from an aspiring management consultant? No: that profession was still nearly a century off. But it was, in effect, the service Rowland Hill had taken it upon himself to perform for Great Britain's postal service.

Hill was a former schoolmaster, whose only experience of the Post Office was as a disgruntled user. Nobody had asked him to come up with a detailed proposal for completely revamping it. He did the research in his spare time, wrote up his analysis, and sent it off privately to the British finance minister, the Chancellor of the Exchequer, naively confident that 'a right understanding of my plan must secure its adoption'.[2]

He was soon to get a lesson in human nature: people whose careers depend on a system, no matter how inefficient

it might be, won't necessarily welcome a total outsider turning up with a meticulously argued diagnosis of its faults
and proposal for improvements. 'Utterly fallacious ... most
preposterous' fulminated the Secretary of the Post Office,
Colonel Maberly; 'wild ... extraordinary' added the Earl of
Lichfield, the Postmaster-General.[3]

Brushed off by the Chancellor, Hill changed tack. He
printed and distributed his proposals, under the title 'Post
Office Reform: Its Importance and Practicability'.[4] He
added a preface, explaining why his very lack of experience
in the postal service qualified him to detect its 'blemishes
and defects'. He wasn't the only person frustrated with the
system, and everyone who read his manifesto – and who
wasn't employed by the Post Office – agreed that it made
perfect sense. The *Spectator* campaigned for Hill's reforms.[5]
There were petitions. The Society for the Diffusion of Useful
Knowledge made representations.[6] Within three years, the
government had bowed to public pressure, and appointed a
Post Office supremo: Rowland Hill himself.[7]

What were the problems Hill identified? Back then, you
didn't pay to send a letter. You paid to receive one. The pricing
formula was complicated and usually prohibitively expensive.
If the postman knocked on your door in Birmingham, say,
with a three-page letter from London, he'd let you read it
only if you coughed up two shillings and threepence.[8] That
wasn't far below the average daily wage,[9] even though 'the
whole missive might not weigh a quarter of an ounce'.[10]

People found workarounds. Members of Parliament could
send letters that would be delivered free of charge – if you
happened to know one, they might 'frank' your letters as a
favour. The free-franking privilege was widely abused – by
the 1830s, MPs were apparently penning an improbable 7
million letters a year.[11] Another common trick was to send

coded messages through small variations in the address. You and I might agree that if you sent me an envelope addressed 'Tim Harford', that would signify you were well; if you addressed it 'Mr. T. Harford', I would understand you needed help. When the postman knocked, I would inspect the envelope, and refuse to pay.

Hill's solution was a bold two-step reform. Senders, not recipients, would be asked to pay for postage; and it would be cheap – one penny, regardless of distance, for letters up to half an ounce. Hill thought it would be worth running the post at a loss, as 'the cheap transmission of letters and other papers . . . would so powerfully stimulate the productive power of the country'.[12] But he made a compelling case that profits would actually go up, because if letters were cheaper to send, people would send more of them.[13]

Economists would recognise the question Hill was trying to answer: how steep was the demand curve? If you reduced the price, by how much would demand increase? Hill didn't know about demand curves: the first such diagram was published in 1838, the year after his proposal.[14] But he knew how to marshal anecdotes: the brother and sister in Reading and Hampstead, some 40 miles apart, who lost touch for three decades, then corresponded frequently when a kindly MP gave them some free-franks.[15] It had only been the expense that put them off.

A few years ago the Indian-born economist C. K. Prahalad argued that there was a fortune to be made by catering to what he called 'the bottom of the Pyramid', the poor and lower-middle class of the developing world. They didn't have a lot of money as individuals, but they had a lot of money when you put them all together. Rowland Hill was more than a century and a half ahead of him. He pointed to a case when small payments from large numbers of poor people had mounted up for the government: duties on 'malt and ardent spirits (which, beyond all doubt, are principally consumed by the poorer classes)' brought in much more than those on 'wine (the beverage of the wealthy)'. Hill concluded, slightly disparagingly:

> The wish to correspond with their friends may not be so strong, or so general, as the desire for fermented liquors, but facts have come to my knowledge tending to show that but for the high rate of postage, many a letter would be written, and many a heart gladdened too, where the revenue and the feelings of friends now suffer alike.[16]

In 1840, the first year of the penny post, the number of letters sent more than doubled. Within ten years, it had doubled again.[17] Hill initially expected that postage-paid envelopes would be more popular than stamps – but the 'Penny Mulready' envelope faded into obscurity, while the 'Penny Black' stamp inspired the world. It took just three years for postage stamps to be introduced in Switzerland and Brazil; a little longer in America; by 1860, ninety countries had them.[18] Hill had shown that the fortune at the bottom of the pyramid was there to be mined.

Cheap postage brought the world some recognisably modern problems: junk mail, scams, and a growing demand for immediate response – half a century on from Hill's penny post, deliveries in London were as frequent as hourly, and replies were expected by 'return of post'.[19]

But did the penny post also diffuse useful knowledge, and stimulate productive power? The economists Daron Acemoğlu, Jacob Moscona and James Robinson recently came up with an ingenious test of this idea in the United States. They gathered data on the spread of post offices in the nineteenth century, and the number of applications for patents from different parts of the country. New post offices did indeed predict more inventiveness, just as Hill would have expected.[20]

Nowadays, what we call 'snail mail' looks to be in terminal decline. There are so many other ways to gladden our friends' hearts. Forms and bank statements are going online; even junk mail is in decline, as spamming us online is more cost-effective: every year, across the developed world, the number of letters sent drops by another few per cent.[21] Meanwhile, the average office worker gets well over a hundred emails a day.[22] We no longer need societies to promote the diffusion of useful knowledge – we need better ways to distil it.

But Acemoğlu and his colleagues think the nineteenth-century postal service has a lesson to teach us today: that 'government policy and institutional design have the power to support technological progress'.[23] What current blemishes and defects in these areas might be holding progress back? We need the successors of Rowland Hill to tell us.

5

Bicycles

One autumn day in 1865, two men sat in a tavern in Ansonia, Connecticut, calming their nerves with a few stiff drinks. They had been riding a wagon down a nearby hill when they heard a blood-curdling scream from behind them. The devil himself, with the head of a man and the body of some unknown creature, was flying down the hill towards them, skimming low over the ground. They whipped their horses and fled, while the devil plunged off the road and into a flooded ditch.

Their fear and awe must have deepened when a dark-haired man who had overheard their story strode across the tavern towards them: bleeding, soaking wet, and French. He introduced himself as the devil.

The devil's real name was Pierre Lallement. The young mechanic had been in the United States for a few months, and had brought with him from France a machine of his own devising – a pedal-cranked, two-wheeled construction he called a velocipede, but which we would call a bicycle. Monsieur Lallement was soon to patent his invention, which still lacked the gears and chain-drive of a modern bicycle. It also lacked brakes – which was why he had plunged down the hill towards the wagoners with such hellish speed.[1]

After a lull of half a century, it was a dramatic rebirth. 'Hobby horses' – two wheels, a seat, no pedals – had been fashionable for a remarkably brief period of time in the summer of 1819, then abandoned as a silly toy. But genuine pedal-cycles? They were about to wreak dramatic changes on the social, technological and perhaps even genetic landscape of the world.[2]

Monsieur Lallement's cumbersome bicycle was soon superseded by the penny-farthing, which was not the genteel vehicle we imagine through the sepia tint of nostalgia. Courtesy of the enormous front wheel, it was a racing machine, twice as fast as a velocipede. It was ridden almost exclusively by fearless young men, perched on top of a five-foot wheel and prone to pitching forward at the slightest obstacle. At which point, explained one cyclist, you'd encounter 'a nice, straight iron handlebar close across your waist to imprison your legs and make quite certain that it should be your face ... that first reached the surface of this unyielding planet.'[3]

But the next technological step, the 'safety bicycle', had much broader appeal. Introduced twenty years after Lallement had swooped downhill like the devil, it looked much like modern bicycles do, with a chain drive, equal-sized wheels, and a diamond frame. Speed came not from a gargantuan wheel, but from gears.[4]

With minor modifications to the crossbar, safety bicycles could even be ridden in a dress. Not that that worried Angeline Allen, who caused a sensation in 1893 by cycling around Newark, on the outskirts of New York City, without one. 'She wore trousers!' bellowed the headline of a popular titillating men's magazine, adding that she was young, pretty, and divorced.[5]

The bicycle was a liberating force for women. Even if they did not emulate Ms Allen's choice of dark blue corduroy bloomers, they needed to shuck off whalebone girdles and hoop-reinforced skirts in favour of something simpler and more comfortable. They would ride without chaperones, too.[6]

The forces of conservatism were alarmed, bellowing that 'immodest bicycling' would lead to masturbation, even prostitution. But these protests soon seemed laughable.

As the cycling historian Margaret Guroff points out, nobody seemed concerned about what Angeline Allen was doing – only what she was wearing while she did it. A woman seen alone in public on a safety bicycle seemed no scandal at all.[7]

Three years later, the elderly Susan B. Anthony, a women's rights activist for most of the nineteenth century, declared that bicycling had 'done more to emancipate woman than any one thing in the world'.[8]

The bicycle continues to empower young women today. In 2006, the state government of Bihar, India, began to heavily subsidise the purchase of bicycles for teenage girls transferring to secondary school – the idea was that the bikes would allow girls to travel several miles to their lessons. The programme seems to have worked, dramatically increasing the chances that girls will stick with secondary school.[9]

Even in America, the bicycle is an inexpensive way to expand horizons: the basketball superstar LeBron James has founded a school that supplies a bike to every student. He says that when he and his friends were on their bikes, they were free. 'We felt like we were on top of the world.'[10]

Yes, the bicycle has long been a liberating technology for the economically downtrodden. In its early days, it was much cheaper than a horse, yet offered the same range and freedom. The geneticist Steve Jones has argued that the invention of the bicycle was the most important event in recent human evolution, because it finally made it easy to meet, marry and mate with someone who lived outside one's immediate community.[11]

But the bicycle ushered in a manufacturing revolution as well as a social one. In the first half of the nineteenth century, precision-engineered interchangeable parts were being used to make military-grade firearms for the US Army, at considerable expense. Interchangeability proved too costly, at first, for civilian factories to emulate fully. It was the bicycle that served as the bridge between high-end military manufacture and widespread mass production of complex products. Bicycle manufacturers developed simple, easily repeatable techniques – such as stamping cold sheet metal into new shapes – to keep costs low without sacrificing quality.[12] They also developed ball bearings, pneumatic tyres, differential gears and brakes.[13]

Both the manufacturing techniques and these innovative components were embraced in due course by auto manufacturers such as Henry Ford. The first safety bicycle was made in 1885 at the Rover factory in Coventry, England. It is not a coincidence that Rover went on to become a major player in the car industry; the progression from making bikes to making cars was obvious.[14]

The bicycle provided stepping stones for modernising Japanese industry, too. The first step was the importing to Tokyo of Western bikes, around 1890. Then it became useful to establish bicycle repair shops. The next step was to begin making spares locally – not too much trouble for a skilled mechanic.[15] Before long, all the ingredients existed to make the bicycles in Tokyo itself – around 1900.[16] By the outbreak of the Second World War, Japan was making more than a million bikes a year, masterminded by a new class of businessman.[17]

It is tempting to view the bicycle as the technology of the past. It created demand for better roads, and allowed manufacturers to hone their skills, and then gave way to the motor car – did it not? The data show otherwise. Half a century ago, world production of bikes and cars was about the same – 20 million each, per year. Production of cars has since tripled, but production of bicycles has increased twice as fast again – to 120 million a year.[18]

And it is not absurd to suggest that bicycles are pointing the way yet again. As we seem to stand on the brink of an age of self-driving cars, many people expect that the vehicle of the future will not be owned, but rented, with the click of a smartphone app. If so, the vehicle of the future is here: globally well over a thousand bike-share schemes and tens of millions of dockless, easy-to-rent bikes are now thought to be in circulation, with numbers growing fast.[19]

Around many gridlocked cities, the bicycle is still the quickest way to get around. Many cyclists are discouraged only by diesel fumes and by the prospect of, like Pierre Lallement, crashing. But if the next generation of automobile is a pollution-free electric model, driven by a cautious and considerate robot, it may be that the bicycle's comeback – just like its first, dramatic, appearance in America – is about to pick up speed.

6

Spectacles

Making spacecraft is not a job at which you can afford to be slapdash. At Lockheed Martin, for example, it used to take a technician two painstaking days to measure 309 locations for fasteners on one curved panel. But now the same job takes little more than two hours, says Shelley Peterson, the aerospace company's head of emerging technologies.[1]

What changed? The technician started wearing glasses. But not just any old glasses: specifically, the Microsoft Hololens. It looks like a bulky set of safety goggles and it layers digital information over the real world – in this case, it scans the curved panel, makes its calculations, and shows the technician exactly where each fastener should go.

Productivity experts are gushing about augmented-reality devices such as the Hololens and Google Glass.[2] When Google debuted their smart glasses in 2012 their prospects seemed quite different.[3] They were seen as a consumer device, something that would let us check Instagram and take videos without the hassle of reaching for our phones. They did not catch on. The few people who ventured out in public wearing Google Glass attracted the dismissive soubriquet 'glassholes'.[4]

Google soon realised their mistake: they'd misidentified their target market. They reinvented their glasses for the workplace. Many jobs, after all, involve frequent pauses to consult a screen that tells us what to do next. With smart specs, we can see those instructions while we keep working. It saves a vital few seconds in getting information from internet to brain.

A thousand years ago, information travelled rather more slowly. In Cairo, in the 1010s, the Basra-born polymath Hasan ibn al-Haytham wrote his masterwork: the *Book of Optics*;[5] it took two centuries for his insights to be translated out of Arabic.[6] Ibn al-Haytham understood vision better than anyone before him. Some earlier scholars, for example, had argued that the act of seeing must involve some kind of rays being emitted from the eye. By careful experiment, Ibn al-Haytham proved them wrong: light comes into the eyes.

Before Ibn al-Haytham, optical devices had been cumbersome: the Roman writer Seneca magnified text using a clear glass bowl of water.[7] But the gradual spread of knowledge inspired new ideas:[8] some time in the late 1200s came the world's first pair of reading glasses. Who made them is lost to history, but they probably lived in northern Italy. Venice, in particular, was a hub of glassmaking at the time – problematically so, as buildings in Venice were made of wood, and the glassmakers' furnaces kept starting fires. In 1291, the city's authorities banished the entire trade to the neighbouring island of Murano.[9]

By 1301, 'eyeglasses for reading' were popular enough to feature in the rulebook of the Guild of Venetian Crystal Workers. But historians' biggest clue to the origin of eyeglasses comes from a sermon in 1306 by one Friar Giordano da Pisa. The invention was now 20 years old, he told his congregation in Florence.[10] It was, he enthused, 'one of the most useful devices in the world'.[11]

He was right. Reading strained the eyes at the best of times: medieval buildings weren't famed for their big windows, and artificial light was dim and expensive.[12] As we age, it gets harder to focus on close-up objects; middle-aged monks and scholars, notaries and merchants, were simply out of luck. Friar Giordano was 50.[13] One can imagine why he appreciated his spectacles so much.

But they were useful only to the small minority who could read. When the printing press came along, glasses reached a bigger market. The first specialist spectacle shop opened, in Strasbourg, in 1466.[14] Manufacturers branched out from convex lenses, which help people to see close-up; they learned how to grind concave lenses, which help people focus on things that are far away.[15]

Put concave and convex lenses together, and you have the basic ingredients for a microscope or a telescope. Both inventions emerged from the spectacle shops of the Netherlands around the year 1600, opening whole new worlds to scientific study.[16]

Nowadays we take glasses for granted – in the developed world, at least. A survey in the UK found that about three-quarters of people wear glasses or contact lenses, or have had surgery to correct their vision.[17] It's a similar story in America and Japan.[18]

In less developed countries, however, the picture is very different – and only recently did we get a clearer view. Historically, the World Health Organization collected data only on how many people have really serious problems with their vision.[19] Many more can see well enough to muddle through daily life, but would still benefit from spectacles. But how many? The world's leading lens-maker, Essilor, decided to find out, no doubt for entirely selfless reasons, and in 2012 came the answer: around the world, 2.5 billion people need glasses and don't have them.[20] That's an eye-popping figure, but serious people think it's credible.[21]

And many of those 2.5 billion may have no idea that glasses could help them. In 2017, researchers went to a tea plantation in Assam. They tested the vision of hundreds of tea-pickers aged 40 or over, and gave a simple $10 pair of reading glasses to half of those who needed them. Then they compared how much tea was picked by those who wore the glasses and those who didn't.

Those with glasses averaged about 20 per cent more tea. The older they were, the more their tea-picking improved. The tea-pickers are paid by how much tea they pick. Before the study, not one owned glasses. By the end, hardly any wanted to give them back.[22]

How widely we can extrapolate from this study is hard to say: picking tea may reward visual acuity more than some other jobs.[23] Still, even conservative estimates put the economic losses from poor eyesight into the hundreds of billions of dollars – and that's before you think about people's quality of life, or children struggling at school.[24] One randomised trial concludes that giving kids glasses could be equivalent to an extra half-year's schooling.[25]

And the need is growing. Presbyopia is long-sightedness that comes with age; but among children there's now a global epidemic of myopia, or short-sightedness. Researchers aren't sure why, though it may have to do with kids spending less time outdoors.[26]

What would it take to correct the world's vision? Clearly, more eye doctors would help – the number varies widely from country to country. Greece, for example, has one ophthalmologist for about five thousand people; in India, it's one per seventy thousand; in some African countries, one in a million.[27]

But while serious eye problems demand skilled professionals, people whose needs are more easily fixable could be reached by other workers. In Rwanda, a charity trained nurses to do sight checks; researchers found they did them well over 90 per cent of the time.[28]

How about teachers? I've worn glasses since primary school, when my teacher saw me squinting at the blackboard and told my mother to take me to an optician. Another study backs up the idea: after just a couple of hours' training, teachers at schools in rural China could identify most children who needed glasses and didn't have them.[29]

It shouldn't be rocket science to roll out thirteenth-century technology. One wonders what Friar Giordano would make of a world in which we build spacecraft in augmented reality, but we haven't yet helped a couple of billion people fix their fuzzy views of actual reality. He'd probably tell us where to focus.

7

Canned Food

Play the word-association game with 'Silicon Valley', and your mind is unlikely to go to 'canned food'. Silicon Valley stands for cutting-edge technology, bold ideas that change the world. Canned food is the height of mundanity: you reach for it when you're too tired, or poor, to cook something interesting. Nobody would say the tin can is cutting-edge technology, although the more literal-minded might make that claim for the can opener.

Yet, in its day, canned food was as revolutionary as anything now being pitched by Bay Area start-ups. And its story reveals how surprisingly little some deep dilemmas around innovation have changed in the last two hundred years or so.

To start with: how do we incentivise good ideas? There's the lure of a patent, of course, or first-mover advantage. But if you really want to encourage fresh thinking, offer a prize. Self-driving cars are a current example. In 2004, the Defense Advanced Research Projects Agency – DARPA – offered a million dollars to the first vehicle to find its way across a course in the Mojave Desert.[1] The result was pure Wacky Races: vehicles caught fire, flipped over, crashed through fences and ground to a halt because they were confused by tumbleweed.[2] But within a decade, self-driving cars were reliable enough to let loose on public roads.[3] Now the technology is a priority for Silicon Valley behemoths from Apple to Google to Uber.

The DARPA prize was hardly the first, however. In 1795, the government of France offered a prize of 12,000 francs for inventing a method of preserving food. It was eventually claimed by Nicolas Appert, a Parisian grocer and confectioner credited with the development of the bouillon cube and, less plausibly, the recipe for Chicken Kiev. Through trial and error Appert found that if you put cooked food in a glass jar, plunged the jar into boiling water, and then sealed it with wax, the food wouldn't go off.[4] Appert had no idea why his method worked – it would be a few decades before Louis Pasteur came along to explain that heat kills bacteria. But it worked. Appert became known as the 'father of canning'.[5]

Why was the French government interested in preserving food? For the same reason DARPA was interested in vehicles that could navigate themselves across deserts: with a view to winning wars. Napoleon Bonaparte was an ambitious general when the prize was announced; by the time it was awarded he was France's emperor, about to launch his disastrous invasion of Russia. Napoleon may or may not have said that 'an army marches on its stomach',[6] but he was clearly keen to broaden his soldiers' provisions from smoked and salted meat.[7]

Appert's laboratory was an early example of an idea we'll encounter often in this book: military needs spur innovations that transform the economy. From GPS to ARPANET, which became the internet, Silicon Valley is built on technologies first funded by the US Department of Defense.

But even when ideas come from the public sector, it takes a culture of entrepreneurship to explore what they can do.

Appert wrote up his experiments; his book was later published in English as *The Art of Preserving All Kinds of Animal and Vegetable Substances for Several Years*, with chapters helpfully devoted to everything from 'New-laid Eggs' to 'Pears of every Kind'.[8] Meanwhile another Frenchman, Philippe de Girard, started applying the techniques to containers made of tin, not glass. But when he wanted to commercialise his idea, he decided to sail across the English Channel.[9]

Why? Too much French bureaucracy, says Reading University's Norman Cowell: 'The philosophy in England was entrepreneurial, there was venture capital. People were prepared to take a risk.' Girard employed an English merchant to patent the idea on his behalf – a necessary subterfuge, as England was at war with Napoleon – and an engineer and serial entrepreneur named Bryan Donkin bought the patent for the tidy sum of £1000. Donkin's factory in Bermondsey was soon supplying everyone from polar explorers to the Duke of Kent.[10]

A modern-day Girard, looking for a place with venture capital and risk-taking attitudes, would surely head for Silicon Valley. For decades, others have tried to emulate its knack for generating ideas and growing businesses – to create an 'innovation ecosystem', in the current parlance.[11] London has its Silicon Roundabout, Dublin its Silicon Docks; Cameroon touts a Silicon Mountain, the Philippines a Silicon Gulf, and Bangalore is less imaginatively dubbed the Silicon Valley of India.[12] But none have yet quite measured up.[13] We economists can confidently tell you some ingredients for an innovation ecosystem, such as making businesses easy to set up and encouraging links with academic research. But nobody has perfected the recipe.

One ingredient that's much debated is how best to regulate. Lack of red tape helped attract Girard to England, but canned food was about to demonstrate why rules and inspections serve a purpose. By 1845, with Donkin's patent now expired, Britain's navy was looking to save some money. They started buying from Stephen Goldner, whose prices were low because labour was cheap where he had his canning factory, in what is now Romania. Unfortunately, that wasn't the only way that Goldner was keeping costs down. After complaints from sailors, naval inspectors started checking his wares: on one occasion they tested 306 cans, and only 42 were edible. The rest contained such delicacies as putrid kidneys, diseased organs and dog tongues.[14]

The scandal hit the newspapers at an unfortunate time, when the Great Exhibition of 1851 had just introduced ordinary Londoners to canned delights hitherto stocked only by luxury foodstores. There were sardines and truffles, artichokes and turtle soup. Putrid kidneys were not supposed to be part of the narrative. With quality improving and prices coming down, canned food had seemed set for the mass market – but it took years to rebuild public confidence.[15]

That mass market seemed self-evidently desirable: with refrigeration yet to be invented, safe canned food would widen people's diets and improve nutrition.[16] It's not always so straightforward to anticipate how new technologies will play out, and whether regulators should try to speed them up or hold them back, to nudge their direction or leave well alone. Take social media: it took barely five years for gushing accounts of how it helped the Arab Spring to give way to hand-wringing about how it helped elect Donald Trump.[17]

Or take the self-driving car. Should we look forward to the convenience, or worry about lost jobs? Will artificial intelligence hugely widen inequality? Should governments step in? How? These are debatable questions, but some Silicon Valley types are concerned enough about where their innovations may be taking us that they're seriously imagining apocalyptic scenarios. We're 'skating on really thin cultural ice right now', says a former Facebook manager to the *New Yorker*, explaining why he's bought land on an island and stocked up on ammunition. Others have bought underground bunkers, and have planes on constant standby in case society implodes. These 'preppers', by one estimate, include at least half the billionaires in Silicon Valley.[18]

Progress can be fragile. Nicolas Appert found that out: he invested his 12,000 francs in expanding his canning operation, only to see it destroyed by invading Prussian and Austrian armies as Napoleon's rule collapsed.[19] The world looks more stable now, and the Silicon Valley preppers are probably worrying too much – but if their worst fears are realised the world's most valuable commodity might yet be . . . canned food.

8

Auctions

In the year 211 BCE, Rome and Carthage were engaged in a long war that was to shape the ancient Mediterranean. The North African general Hannibal had vanquished Roman legions at will. As the Romans regrouped and began to fight back, Hannibal decided on a bold feint: he would march on Rome itself. Although he had little hope of breaching the city's defences, he hoped the Romans would panic and recall their armies. The historian Edward Gibbon relates what happened next:

> He encamped on the banks of the Anio, at the distance of three miles from the city; and he was soon informed, that the ground on which he had pitched his tent, was sold for an adequate price at a public auction.[1]

The implication was obvious: Rome had seen through the bluff. If Romans were willing to trade at full price the land underneath Hannibal's army, they did not expect his army to linger. It did not: Hannibal withdrew in short order.

This may be the only example of an auction being used as an

attack on enemy morale, but it is not the first recorded auction. For example, three hundred years earlier Herodotus describes men bidding for the most attractive wives in Babylon:

> The rich men who wanted wives bid against each other for the prettiest girls, while the humbler folk, who had no use for good looks in a wife, were actually paid to take the ugly ones.[2]

Problematic, yes. But ingenious: this auction was a community affair in which funds raised from the high bidders were used to compensate the poorer men.

Auctions seem to be almost as common as the marketplace itself. You can imagine the idea being endlessly rediscovered around the world, whenever some trader offered to pay three obols per jar for a shipment of olive oil, and the man next to him said, 'Don't take that offer – I'll pay four.'

From such simple moments evolved the theatrical event we called the open-outcry auction – a room full of art or antique dealers, millionaire backers submitting bids by phone, and a dapper auctioneer tickling the whole process along. Going once, going twice, gone!

By making clear what others are prepared to pay, such auctions make it hard for the unscrupulous to exploit the gullible. In the early nineteenth century, British traders used auctions to offload large volumes of inexpensive British products in the United States. American consumers were delighted; American merchants were indignant.[3] One of them, Henry Niles, complained in 1828:

> [Auctions are] the grand machine by which British agents at once destroy all regularity in the business of American merchants and manufacturers.[4]

An anti-auction committee lobbied Congress, declaring:

Auctions are a monopoly; and like all monopolies are unjust, by giving to a few, that which ought to be distributed among the mercantile community generally.[5]

That was special pleading – the 'mercantile community' just wanted to preserve their mark-ups. Yet there is an important grain of truth in the complaint: in any auction, the sellers want to be where the buyers are, and the buyers want to be where the sellers are. That makes auctioneering a natural monopoly: there is always a risk that large auction venues abuse their market power.

While the open-outcry auction is the most famous kind, there are many other ways to design an auction. The seventeenth-century diarist Samuel Pepys describes an auction 'by an inch of candle', which ended when the flame of the candle stub flickered out. The unpredictability of this moment was intended to prevent people from using the unpopular tactic of submitting a bid at the last second.[6]

If not by candle, what about an auction by clock? The 'Dutch clock auction' is used at the vast flower market of Aalsmeer, and the clock face shows not the time, but the price. That price ticks down and down, until somebody stops the clock by pressing a button. Whoever stopped the clock buys the flowers at the price specified. At first glance, the method could hardly be more different from an open-outcry auction. The fundamentals are not so very different, though – and they are even faster, as befits a product that will wilt fast if it cannot be sold and shipped.[7]

Then there is the sealed-bid auction, beloved of estate agents. Write down your bid, slip it into an envelope, and seal it tight. Highest bid wins the prize. But here's a curiosity:

under the surface, the sealed-bid auction is exactly the same as the Dutch flower-clock auction. In each case, you simply need to pick your price. Unlike in an open-outcry auction, you'll learn nothing about anybody else's willingness to pay until it's too late.

The Nobel laureate economist William Vickrey produced a famous theorem demonstrating that under ideal conditions, all auctions can be expected to raise the same amount of revenue.[8] Like any economic theorem, that oversimplifies the case. Auction details can matter a lot – if an auction opens up a loophole for cheats, or discourages bidders from bothering, it can fail badly.[9]

One might ask why auctions are used in some circumstances, while in other cases sellers post a take-it-or-leave-it price. Your local supermarket, for example, does not auction off the cabbages.

The answer is that auctions come into their own when nobody is quite sure of the value of what is being sold. Second-hand products sold on eBay are an obvious example, but there are many others – a permit to drill for oil in unexplored terrain; a painting by Leonardo da Vinci; a licence to use radio spectrum to provide mobile phone services . . . This common resource – the radio spectrum – used to be handed out to favoured companies for trivial sums. Now governments auction it off for billions.[10]

In each case, the true value is unknown. Each bidder will have their own information. The auction brings together all of that information, and transforms it into a price. It is quite a trick. And it is something the Romans understood when they reported the results of an auction to tell Hannibal: we're not scared.

While auctions seem reassuringly old-school, they take place at the cutting edge of the modern digital economy.

Think about what happens when you type a search term into Google. Alongside your search results, you'll see advertisements. Those adverts are there because they won a complex auction which assigns them more or less prominent positions depending both on their bid-per-click, and on how good the Google algorithm thinks the advertisement is.[11]

For example, an art dealer might offer a high fee to appear next to searches for 'Picasso', but an advertiser selling Picasso posters might expect many more clicks – and win top spot in the auction for a lower per-click bid.

These auctions take place every time someone types a search into Google, and their scale is unnerving: Google's parent company Alphabet makes more than \$2 billion profit every month.[12] Most of that is from advertising, and most of the advertising is sold by auction. In 2019 Google was estimated to have taken more revenue from advertising than its two biggest rivals – Facebook and Alibaba – combined.[13]

Often you see an advert for Google's own products. Is it a problem that Google bids in its own auctions? It is hard to be sure. You can imagine how any company might benefit from intimate knowledge of its rivals' strategies in bidding for ad space, though Google insists it gets no unfair advantage from its dominant market position.[14]

Henry Niles, the anti-auction activist, would no doubt have had something to say about that.

II

SELLING THE DREAM

9

Tulips

One frosty winter morning, at the start of 1637, a sailor presented himself at the counting house of a wealthy Dutch merchant and was offered a hearty breakfast of fine red herring. The sailor noticed an onion – or so he thought – lying on the counter. Here is what happened next, according to Charles Mackay, writing in Scotland two centuries later:

'Thinking it, no doubt, very much out of its place among silks and velvets, he slyly seized an opportunity and slipped it into his pocket, as a relish for his herring. He got clear off with his prize, and proceeded to the quay to eat his breakfast.'

Alas, says Mackay, 'Hardly was his back turned when the merchant missed his valuable *Semper Augustus*, worth three thousand florins, or about 280 pounds sterling.'[1]

Relative to the wages of the time, that is well over a million dollars today. Seeking a zesty accompaniment to his fish, the sailor had unwittingly pilfered not an onion, but a rare *Semper Augustus* tulip bulb. And in early 1637, tulip bulbs were reaching some truly extraordinary prices.

Then, very suddenly, it was over: in February, bulb wholesalers gathered in Haarlem, a day's walk west of Amsterdam, to find that nobody wished to buy. Within a few days, Dutch tulip prices had fallen tenfold.[2]

Tulip mania is often cited as the classic example of a financial bubble: when the price of something goes up and up, not because of its intrinsic value, but because people who buy it expect to be able to sell it again at a profit. It seems foolish to pay a million dollars for a tulip bulb – but if you hope to sell it on to a greater fool for two million, it can still be a rational investment. This is known as 'greater fool' theory.

Whether or not it explains tulip mania is, however, a subtle question.

Charles Mackay's 1841 account has cast a long shadow over our imagination. His book *Extraordinary Popular Delusions and the Madness of Crowds* is full of vivid stories about how the entire Dutch nation was involved. But those extravagant tales – including the one I have just told you, about the hungry sailor – are probably false.

Tulips were part of a cornucopia of new plants to arrive in Europe in the sixteenth century: potatoes, green and red peppers, tomatoes, Jerusalem artichokes, French beans, runner beans. At first, tulip bulbs were sufficiently unfamiliar to be mistaken for vegetables: on at least one occasion, someone roasted them with oil and vinegar – the germ of truth in Charles Mackay's tall tale.[3]

But once it became clear what to do with tulips, soon everyone was waxing lyrical about their beauty. Some, infected by a virus, changed from simple bold-coloured petals to exquisitely varied patterns. Just as the super-rich today collect beautiful paintings at extraordinary prices, the newly wealthy Dutch merchant class began to collect and display rare tulips.

And not always honestly. The celebrated botanist Carolus Clusius generously shared his tulips with friends and colleagues, yet suffered many thefts of rare plants. His treasures were, after all, just sitting in gardens. Once, Clusius had some unique flowers stolen, only to find them in the garden of a Viennese aristocrat. She, of course, denied all knowledge.[4]

The philosopher Justus Lipsius wasn't impressed by the tulip collectors. 'What should I call this but a kinde of merrie madnesse?' he said, adding, 'They do vaingloriously hunt after strange herbs and flowers, which having gotten, they preserve and cherish more carefully than any mother doth her childe.'[5]

But, in the early 1600s, the price of tulips just kept on rising. Adriaan Pauw, who was fabulously wealthy and the closest thing Holland had to a prime minister at the time, built a garden full of artfully positioned mirrors. In the centre stood a few rare tulips, made by the mirrors to look like a multitude. It was an admission that not even Pauw could afford to fill his garden.[6]

The highest price for which we have good evidence was 5200 guilders for a single bulb, in that winter of 1637. That is more than three times what Rembrandt charged for painting *The Night Watch* just five years later, and twenty times the annual income of a skilled worker such as a carpenter. The idea that some poor fellow had his million-dollar tulip bulb consumed with a herring may be fanciful; the idea that the rarest bulbs were million-dollar treasures is about right.[7]

Could a tulip bulb really be worth a million dollars? It is not quite so absurd as it might seem. Tulip bulbs produce not only tulips, but offshoot bulbs called offsets. If the tulip has a beautiful pattern the offsets will also tend to produce such patterns. Owning a rare bulb was a bit like owning a champion racehorse: valuable in its own right, perhaps, but far more valuable because of its potential offspring.[8] Given how far the wealthy would go to possess unusual tulips, there was nothing foolish about bulb traders paying top guilder for the bulbs.

Financial bubbles burst when expectations reach a tipping point: once enough people expect prices to fall, the supply of greater fools dries up. Does that explain the sudden collapse in prices in February 1637? Perhaps.

But there's another theory. As rare bulbs such as *Semper Augustus* multiplied over the years, it is only natural that their price would fall.[9] In Haarlem – one of the warmer Dutch cities – February is exactly when tulip shoots would have burst through the soil. Having seen abundant shoots on their journeys, the bulb traders might have realised that the crop would be bountiful, and the rare flowers rather less rare than they had imagined.[10] If so, the fall in prices may have reflected an increase in supply, rather than the bursting of a bubble.

Whatever the reason, the mania subsided. The fallout was painful: many trades were not simple exchanges of cash for bulbs, but promises to pay for bulbs in future. Between buyers who didn't have money and sellers who didn't have bulbs, there was a good deal of grumbling over who owed what to whom. But the prosperous Dutch economy sailed on regardless.

Later bubbles were much more consequential. Perhaps the greatest boom and crash in history was the railway mania of the 1840s. Influential commentators waved away warnings of financial trouble ahead, and encouraged investors to bid up stocks in UK railway companies to absurd prices.

One such bullish commentator was Charles Mackay, who had revelled in telling those wildly inaccurate but deeply satisfying stories about the greed and foolishness of the Dutch tulip speculators. Mackay's book about collective madness came out in a new edition after the railway bubble burst. Oddly enough, he said very little about it.[11]

It's easy to scoff at past bubbles; it is not so easy to know when one may – or may not – be in one.

10

Queen's Ware

'To this manufacture the Queen was pleased to give her name and patronage, commanding it to be called Queen's Ware, and honouring the inventor by appointing him Her Majesty's potter.'

At least, that was Josiah Wedgwood's story. Wedgwood's biographer, Brian Dolan, reckons it's more likely that Queen Charlotte's 'command' was Wedgwood's suggestion.[1] She probably saw it more as flattery than shrewd self-interest.

Why more likely? Because Josiah Wedgwood was a shrewd individual. He was perhaps the world's first management accountant, and a pioneering early chemist, endlessly experimenting with new ways to treat and fire clay, and noting his results in a secret code lest a rival steal his notebook. His first big breakthrough was the new kind of 'creamware', or cream-coloured pottery, from which he'd fashioned the tea service that so impressed the Queen: 'quite new in its appearance,' he noted modestly, 'covered with a rich and brilliant glaze'.[2]

Wedgwood was a lobbyist. In the 1760s, North Staffordshire potters had to dispatch their fragile wares over miles of bone-shaking, pot-breaking roads to get to major cities.[3] Wedgwood roused investors and persuaded parliament to approve a canal connecting the Trent and the Mersey. His fellow potters were delighted, until they realised that Josiah had cannily snapped up land and built his enormous new factory right on the banks of where the canal would pass.[4]

But perhaps Josiah's most impressive achievement was solving a problem in monopoly theory two centuries before it was even articulated.

The man who put the problem into words was a Nobel memorial prize-winning economist called Ronald Coase. Imagine, said Coase, that you're a monopolist – you alone produce a certain good. Many people want to buy it; some would pay a lot, others much less, although still enough for you to turn a profit. Ideally, you'd like to charge a high price to the first group, a low price to the second.

But how can you get away with that? One possible answer is to launch at a high price, then lower it to widen your market. That's what Steve Jobs tried with the first iPhone. It cost $600. After two months, he cut the price to $400. Not surprisingly – although it surprised Steve Jobs – the people who'd rushed to pay $600 were less than impressed.[5]

Coase said this strategy can't work. The first set of buyers will see through the trick. They'll realise that if they only wait, they can get the good more cheaply. This idea is called the Coase Conjecture, and Coase published his paper explaining it in 1972.[6]

Back in 1772, Josiah was putting into words the business model that had taken shape in his mind since his meeting with the Queen and his dabbles in management accounting. He'd grasped the difference between what economists now call fixed costs, such as research and development, and variable costs, such as labour and raw materials.[7] It initially incurred a *'great price'*, he mused to his business partner, to 'make the Vases esteemed *Ornaments for Palaces'*.

But once he'd perfected the process and trained his workers, he could churn out copies cheaply. By this time, 'The Great People have had their Vases in their Palaces long enough for them to be seen & admir'd by the *Middling Class* of People.' You can almost hear the cash registers pinging as Josiah writes on: 'the middling People would probably buy quantities of them at a reduced price'.[8]

Josiah had anticipated what later became known as the 'trickle-down' theory of fashion: people tend to emulate those they consider to be above them on the social scale.[9] It's not the only theory – nowadays trends also 'trickle up', from the cool kids on the street;[10] fashion companies employ 'coolhunters' to identify them.[11]

But trickle-down still works. Why else, for example, would the jeweller Anna Hu reportedly pay the actress Gwyneth Paltrow a million dollars to wear her diamond bracelet to the Oscars?[12] She must have hoped to recoup the cost by inspiring purchases from the 'middling People'.

Before we had movie royalty there was only, well, royalty: in the 1760s you couldn't get much higher on the social scale than the Queen of England. Josiah's 'Queen's Ware' gambit worked spectacularly. Sales were 'really amazing', he wrote. Queen's Ware sold at twice the price of rivals' comparable goods: as the historian Nancy Koehl puts it, 'The middling-class customers were to be won with quality and fashionability, rather than low prices.'[13]

Josiah asked himself the key question: 'How much of this general use & estimation is owing to the mode of its introduction & how much to its real utility & beauty?' From now on, he concluded, he should bestow 'as much pains & expence' on gaining 'Royal or Noble' approval for his products as on the products themselves.[14]

But what should Josiah make next? He set off to do some coolhunting of his own. He courted the 'virtuosi' – wealthy art collectors who brought back pieces from their Grand Tours in Europe. The hottest new thing, he discovered, was the Etruscan pottery now being excavated in Italy.[15] Could Josiah make something similar? He got to work in his laboratory – bronze powder, vitriol of iron, crude antimony – and concocted a pigment that let him imitate the Etruscan style to perfection.[16] In a shameless piece of branding, he called his canal-side factory 'Etruria'.

Aristocratic clients lapped it up: you shall 'exceed the Antients', gushed one elderly Lord, ordering three vases.[17]

And Josiah kept experimenting. Traditionally, clay was fired and then painted or enamelled. He figured out how to dye the clay with metal oxides before firing it, producing an oddly translucent effect. 'Jasper ware' came in a distinctive light blue, with white decorations in relief, that's still associated with the Wedgwood brand.[18]

It was another huge success. In the words of the historian Jenny Uglow, Josiah could 'not just follow the fashion but set it'.[19]

But why did Wedgwood not fall foul of the Coase Conjecture? After a while, his aristocratic clients must have figured out that whenever Wedgwood launched something they'd never seen before, they could simply wait to pick it up more cheaply.

The answer lies in the trickle-down theory of fashion: if people are trying to emulate their social superiors, what do you do if you're already at the top of the scale? You try, of course, to look different to the people below you. Some economists now analyse fashion as an exception to the Coase

Conjecture.[20] Even if you know you'll get something cheaper if you wait a while, sometimes you still want it right now.

A few years after he wowed the Queen, Josiah observed that Queen's Ware was 'now being render'd vulgar & common everywhere'.[21] If the Great People wanted to set themselves apart from the Middling People, they'd have to show off their wealth and good taste by buying something new. Josiah Wedgwood always had something new to sell them.

11

The Bonsack Machine

Camel cigarettes are 'terrible and stick in one's throat', said one man in a blind taste-test in 1920s America. That's how he knew the cigarette he was smoking must be a Lucky Strike, his usual brand. Luckies, you see, 'go down easy and smooth', just like this one.

He was, of course, smoking a Camel.[1]

Nowadays, the awesome power of branding is hardly news. Back then, it was only just beginning to become apparent. Early big-name brands included Kellogg's cereal, Campbell's soup and Colgate toothpaste,[2] but nowhere was branding more crucial than with cigarettes. Unprecedented sums went into launching Camels, for instance: in 1914, day after day, newspaper ads built excitement – 'The Camels are coming'; then 'Camel cigarettes ARE HERE'.[3] The historian Robert Proctor, author of *Golden Holocaust*, reckons 'it is probably fair to say that the industry *invented* much of modern marketing'.[4]

Why did cigarettes lead the way?

There's never a single answer. Cigarettes might have struggled without the accidental discovery of flue-curing in 1839, which made tobacco less alkaline. That meant you could suck smoke into your lungs, which is more addictive than holding it in your mouth. The invention of the safety match helped, too.[5] But the starring role goes to an inventor from Virginia called James Bonsack.

In 1881, when Bonsack patented his new machine, tobacco had been around for centuries but cigarettes remained a niche product: the market was dominated by pipes, cigars and chewing tobacco. Bonsack's father owned a wool factory. He looked at the factory's carding machine – carding is a step in turning fibres into yarn – and wondered if he could adapt it for rolling cigarettes. The contraption he came up with weighed a ton. It churned out two hundred cigarettes a minute – almost as many as a human, rolling by hand, could make in an hour.[6]

The significance was clear to the tobacco entrepreneur James Buchanan 'Buck' Duke, who promptly cut a deal with Bonsack and set about cornering the cigarette market. But Duke's opportunity was also a challenge: he could make lots of cigarettes – but could he sell them? Cigarettes had an image problem: they were seen as lower-status than cigars, which – crucially – were proving altogether harder to mechanise.[7]

Duke wasn't daunted. He saw what he had to do: advertise. He came up with gimmicks like coupons and collectable cards. By 1889, he was spending some 20 per cent of his revenues on promotion – unheard of at the time.[8] And it worked. By 1923, cigarettes had become the most popular way for Americans to consume tobacco.[9]

Some early ad campaigns would now raise an eyebrow. Lucky Strikes, for instance, were pitched as an aid to slimming: 'Reach for a Lucky instead of a sweet' went the tagline, accompanied by an image of a svelte young lady. Sweet makers were outraged. One advertised back, 'Do not let anyone tell you that a cigarette can take the place of a piece of candy. The cigarette will inflame your tonsils, poison with nicotine every organ of your body, and dry up your blood – nails in your coffin'.[10]

But who would you trust for health advice: candy companies or medical professionals? '20,679 Physicians say "Luckies are less irritating"', said a Lucky Strike campaign. If that wasn't enough physicians for you, how about 'More doctors smoke Camels than any other cigarette'.[11]

The blind taste-tests suggested that those claims about throat irritation were spurious, and in the 1940s a more systematic investigation by *Reader's Digest* magazine reached the same conclusion: in terms of health, it makes 'no earthly difference' which brand you buy.[12] In the 1950s, American regulators decided they shouldn't allow cigarette adverts to reference doctors or body parts.[13] It looked like a crisis for the advertisers, but it turned out to be liberating – a realisation dramatised in the television series *Mad Men*:

> This is the greatest advertising opportunity since the invention of cereal. We have six identical companies making six identical products. We can say anything we want.[14]

The ad man – Don Draper – was fictional, but his insight was on point. When products are essentially indistinguishable, companies might compete on price – but that will erode their profit margins. Much better to compete on branding: make people think the products are different, so you can appeal more effectively to different buyers. In the 1960s, Americans bought more cigarettes than ever.[15] Perhaps you'd smoke Marlboro to associate yourself with the rugged masculinity of the Marlboro Man; perhaps you'd see the slogan 'You've Come A Long Way, Baby', and signal your approval of feminism by smoking Virginia Slims.[16]

Economists talk about the 'consumer surplus' produced by a product – that's the enjoyment the product produces, less the money you have to give up to afford it. Does it matter if that enjoyment comes from your appreciation of a product's qualities, or your fond beliefs about the brand? In other words: if you confidently misidentify Camels as Lucky Strikes in a blind taste-test, should we take less seriously the enjoyment you say you get from Lucky Strikes?

No doubt we can be relaxed about this question when it comes to cornflakes, or soup, or toothpaste. If you're swayed by ads for Kellogg's or Campbell's or Colgate, where's the harm? But with a product as deadly as cigarettes, we might worry that the consumer experience is bound up with the branding. Many countries have duly banned television adverts and sports sponsorship.[17] Some even insist on plain packaging, with brand names rendered in a boringly uniform typeface.[18] Tobacco companies say there's 'no compelling evidence' this works.[19] And that might be more persuasive if they hadn't spent years saying there was no compelling evidence that cigarettes caused cancer or heart disease.[20]

In many places, smoking is now in decline; but in some poorer countries with looser regulations, it's a different story.[21] Around the world, still about 6 trillion cigarettes are made every year. That is more than a thousand for each adult on the planet.[22]

For example, in China, in the half-century after Chairman Mao took power, per capita cigarette sales went up roughly tenfold.[23] The China National Tobacco Corporation is the country's most profitable company, and it sells 98 per cent of cigarettes. It's state-owned, and contributes up to a tenth of government revenues. Perhaps it's no surprise, then, that China has been late to the game with restrictions on advertising. As recently as 2005, adverts assured that: 'Smoking removes your troubles and worries'. One brand warned that: 'Quitting smoking would bring you misery, shortening your life'. That brand's name? Longlife.[24]

Soon after, the China National Tobacco Corporation embarked on a new policy: 'premiumisation'. China was getting richer, so why not persuade consumers to pay more for their smokes? It launched new 'premium' brands, which adverts touted as less harmful, higher-quality and more prestigious for gift-giving. And it worked. Before, discount brands outsold premium brands ten-to-one. After nine years, they were near parity.[25]

According to one study, just 10 per cent of smokers in China are aware that brands labelled 'light-' and 'low-tar' are no less harmful to your health than other cigarettes.[26] It seems they haven't read the *Reader's Digest* – and the power of the brand to create credulity is still as strong as ever.

12

The Sewing Machine

Gillette ads stand against toxic masculinity.[1] Budweiser makes specially decorated cups to encourage non-binary and genderfluid people to feel pride in their identity.[2] These examples of 'woke capitalism', of corporations promoting progressive social causes, feel ostentatiously up-to-the-moment.[3] But woke capitalism isn't as new as you might think.

Back in 1850, social progress had further to go. A couple of years earlier, the American campaigner Elizabeth Cady Stanton had caused controversy at a women's rights convention by calling for women to get the vote. Even her supporters worried that that was too ambitious.[4]

Meanwhile, in Boston, a failed actor was trying to make his fortune as an inventor. He'd rented space in a workshop showroom, hoping to sell his machine for carving wooden type. But wooden type was falling out of fashion. The device was ingenious, but nobody wanted to buy one.[5]

The workshop's owner invited the inventor to take a look at another product he was struggling with. It was a sewing machine, and it didn't work very well. Nobody had yet succeeded in making one that did, though inventors had been trying for decades.

The opportunity was clear. True, the time of a seamstress was not expensive: as the *New York Herald* said, 'We know of no class of workwomen who are more poorly paid for their work or who suffer more privation and hardship.'[6] But sewing took so much time – 14 hours for a single shirt[7] – there'd be a fortune in speeding it up.

And it wasn't only seamstresses who suffered: most wives and daughters were expected to sew. This 'never-ending, ever-beginning' task, in the words of the contemporary writer Sarah Hale, made their lives 'nothing but a dull round of everlasting toil'. Young ladies had 'busy fingers and vacant minds'.[8]

In that Boston workshop, the inventor sized up the machine, and quipped, 'You want to do away with the only thing that keeps women quiet.'[9]

This man was Isaac Merritt Singer. He was flamboyant, charismatic, capable of great generosity – but ruthless, too. He was an incorrigible womaniser who fathered at least 22 children. For years he managed to run three families, not all of whom knew that the others existed, and all while technically still married to someone else entirely. At least one woman complained that he beat her.

Singer was, in short, not a natural supporter of women's rights – although his behaviour might have rallied some women to the cause. His biographer, Ruth Brandon, dryly remarks that he was 'the kind of man who adds a certain backbone of solidity to the feminist movement'.[10]

Singer contemplated the machine. 'In place of the shuttle going round in a circle,' he told the workshop owner, 'I would have it move to and fro in a straight line, and in place of the needle bar pushing a curved needle horizontally,

I would have a straight needle moving up and down.'[11] Singer patented his tweaks, and started to sell his version of the machine. It was impressive: the first design that really worked. You could make a shirt in just an hour.[12]

Unfortunately, it also relied on various other innovations which had already been patented by other inventors – such as the grooved, eye-pointed needle, to make a lock stitch, and the mechanism for feeding the cloth.[13] In the 'sewing-machine war' of the 1850s, rival manufacturers showed more interest in suing each other for patent infringement than in selling sewing machines.[14] This kind of situation is now known as a 'patent thicket'.

Finally, a lawyer banged heads together: between them, he pointed out, four lots of people owned patents to all the elements you needed for a good machine. Why not license each other, and work together to sue everyone else?[15] It was a stroke of genius: 'patent pools' are now common for complex inventions.[16]

Freed from legal distractions, the sewing-machine market took off – and Singer came to dominate it. That might have surprised anyone who'd seen the rivals' factories. Others rushed to embrace what was known as the 'American system' of manufacturing, using bespoke tools and interchangeable parts. This system had proved its worth in gunmaking, as we'll see. Yet Singer was late to the party: for years his machines were made from hand-filed parts and store-bought nuts and bolts.[17]

But Singer and his canny business partner, Edward Clark, were pioneers in another way: marketing. Sewing machines were expensive, costing several months' income for the average family.[18] Clark came up with the idea of hire purchase: families could rent the machine for a few dollars a month – and, when their rental payments totalled the purchase price, they'd own it.[19]

This was an attractive proposition: no debt, no obligation to buy.[20] It helped overcome the bad reputation built up by the slower, less reliable machines of bygone years. So did Singer's army of agents, who would set up the machine when you bought it, and call back to check that it was working.[21] Singer employed these representatives all around the world: the 'Herald of Civilization', as the company immodestly proclaimed.[22]

Still, all these marketing efforts faced a problem. And that problem was misogyny.

For a flavour of the attitudes Elizabeth Cady Stanton was up against, consider two cartoons. One shows a man asking why buy a 'sewing machine' when you can marry one; in another, a salesman says women will get more time to 'Improve their Intellects!' The absurdity was understood.[23] Such prejudice fuelled doubts that women could operate these expensive machines.[24]

Singer's business depended on showing that they could, no matter how little respect he might have shown for the women in his own life. He rented a shop window on Broadway, and employed young women to demonstrate his machines – they drew quite a crowd.[25] He took them on tour to fairs and carnivals, where he indulged his theatrical impulses by belting out a dolorous popular song about the lot of the seamstress:[26]

> With fingers weary and worn,
> With eyelids heavy and red,
> A woman sat, in unwomanly rags,
> Plying her needle and thread,
> Stitch! Stitch! Stitch!

Singer's adverts cast women as decision-makers: 'Sold only by the Maker Directly to the Women of the Family'.[27] They implied that women should aspire to financial independence: 'any good female operator can earn with them ONE THOUSAND DOLLARS A YEAR'.[28]

By 1860, the *New York Times* was gushing: no other invention had brought 'so great a relief for our mothers and daughters'; seamstresses had found 'better remuneration and lighter toil'.[29] Still, the *Times* rather undercut its gender-conscious credentials by attributing all this to the 'inventive genius of man'. Perhaps we should ask a woman. Here's Sarah Hale, in *Godey's Lady's Book and Magazine*: 'the needlewoman is ... able to rest at night, and have time through the day for family occupations and enjoyments. Is this not a great gain for the world?'[30]

There are plenty of sceptics about 'woke capitalism' today. It's all just a ruse to sell more beer and razors, isn't it? Perhaps it is. Isaac Singer liked to say he cared only for the dimes.[31] But he also showed that social progress can be advanced by the most self-interested of motives.

13

The Mail-Order Catalogue

'B eware! Don't Patronize "Montgomery, Ward & Co."
They Are Dead-Beats'.

Because the Montgomery Ward brand still exists,[1] I'd
better make clear that's not my current advice. It was a warn-
ing issued by the *Chicago Tribune* on 8 November 1873.[2]

What had Aaron Montgomery Ward done to convince
the *Tribune*'s editorial staff that he was running a 'swindling
firm' preying on 'gulls' and 'dupes' in the countryside? Well,
Ward's flyers were offering suspiciously 'Utopian' prices on
a suspiciously wide range of goods: over two hundred items.
And – get this – not only did Montgomery, Ward & Co. not
display their wares in a shop, they employed no agents: 'in
fact they keep altogether retired from the public gaze, and are
only to be reached through correspondence sent to a certain
box in the Post Office'.[3]

Obviously, this must be some sort of scam. Mustn't it?

It seems not to have occurred to the *Tribune* that Ward
might be able to offer his 'Utopian' prices precisely because
he kept no expensive premises and employed no middle men.
But the threat of a lawsuit soon helped the editors to wrap

their heads around Ward's new business model, and a few weeks later they printed a grovelling apology. This was, they admitted, 'a bona fide firm, composed of respectable persons and doing a perfectly legitimate business in a perfectly legitimate manner'.[4] Ward printed it on his next flyer.[5]

Still in his twenties, Aaron Montgomery Ward had made his way to Chicago after clerking at a country store, and got a job as a salesman for the future department store magnate Marshall Field. That job involved touring more of those rural general stores in farming communities, and Ward realised how limited was the selection of goods they stocked, and how high were the prices.[6]

The farmers had noticed that, too. They were already exploring alternative ways to get goods more cheaply to their far-flung rural outposts. They clubbed together in local chapters of a recently formed organisation with a mouthful of a name: the National Grange of the Order of Patrons of Husbandry, or 'the Grange' for short. The hope was that by pooling their buying power, they might negotiate better prices.[7] It's a well-worn idea; its modern incarnations include Groupon.

Mail order existed, but it wasn't common – just a few specialist firms with a limited line of wares.[8] The opening Ward saw was ambitious but simple: use mail order to sell many things, directly, with low mark-ups on wholesale prices.[9] And buyers would pay on delivery – so if they didn't like what was delivered, they could refuse to pay and send it back. It was, as a chastened *Chicago Tribune* admitted, 'difficult to see how any person can be swindled or imposed upon by business thus transacted'.[10] Ward, whose talents extended to copywriting, later gave the world the enduring phrase 'satisfaction guaranteed or your money back'.[11]

Just two years after rousing the *Tribune*'s suspicions, Ward's flyer had become a 72-page catalogue, listing nigh on two thousand items.[12] You could, for example, buy 250 five-inch canary-coloured envelopes for 55 cents; or the same sum would get you twelve dozen small kerosene wicks. For $6.50, you could treat yourself to an extra-fine white woollen blanket, large size. Ward printed testimonials from satisfied customers, some mentioning that he'd undercut their local store by half.[13]

It was basically just a list of goods and prices, but Ward's catalogue was later named by a New York literary society, the Grolier Club, as among the hundred most influential books in American history, up there with *Moby-Dick*, *Uncle Tom's Cabin* and *The Whole Booke of Psalmes*.[14] It was, they said, 'perhaps the greatest single influence in increasing the standard of American middle-class living'.[15]

And it inspired competitors – notably Sears Roebuck, which soon became the market leader. (The story goes that the Sears Roebuck catalogue had slightly smaller pages than Montgomery Ward's – with the intention that a tidy-minded housewife would naturally stack the two with the Sears catalogue on top.)[16]

By the century's end, mail-order companies were bringing in $30 million a year – a billion-dollar business in today's terms;[17] in the next twenty years, that figure grew almost twenty-fold.[18] The popularity of mail order helped fuel demands to improve the postal service in the countryside – if you lived in a city, you'd get letters delivered to your door, but rural dwellers had to schlep to their nearest post office. The government gave in – and realised that if they were going to send postal workers into the boondocks, they'd better improve the road network, too.[19]

'Rural free delivery' was a huge success. Montgomery Ward and Sears Roebuck were among the main beneficiaries.[20] This was the golden age of mail order. Catalogues ballooned to a thousand lavishly illustrated pages.[21] New editions were eagerly awaited. Forget canary-coloured envelopes – you could buy an entire house. For $892, for instance, Sears Roebuck would send you a five-room bungalow. Strictly speaking, they'd send you 'Lumber, Lath, Shingles, Mill Work, Flooring, Ceiling, Finishing Lumber, Building Paper, Pipe, Gutter, Sash Weights, Hardware and Painting Material'.[22] Oh, and plans, which must have been more daunting than the ones you get from Ikea for a Billy Bookcase. Many of these mail-order kit houses are still standing, a hundred years on; some have changed hands for over a million dollars.[23]

The catalogue itself has endured less well. Montgomery Ward and Sears both started building department stores – as wider car ownership made mall shopping more popular, the catalogues became irrelevant. Montgomery Ward nixed theirs in 1985,[24] Sears a few years later.[25] Then came the internet: Jeff Bezos feels no need to send you a thousand-page Amazon catalogue every year. And companies have other ways to inform customers about their products: marketing gurus will tell you that emails get far fewer responses than snail-mailed glossy catalogues, but because they're so much cheaper, they can still bring a higher return on investment.[26]

But if the heyday of the mail-order catalogue has long since passed, its dynamics are now playing out all over again: the world's rising economic power; a government building roads and communications infrastructure in isolated rural areas;[27] customers fed up with existing retail options;[28] visionary entrepreneurs with new business models that let you browse and order from home.

The country is China. For the postal service, read the internet. And the roles of the mail-order giants are being played by China's e-commerce giants – JD.com and Alibaba.[29]

China has thrown itself into online shopping: its citizens spend about as much online as those of the US, the UK, France, Germany and Japan – combined.[30] And drawing rural areas into the economy isn't just about expanding consumer choice and middle-class living standards: when you have good roads and access to information, you have more scope to make and sell stuff. The economists James Feigenbaum and Martin Rotemberg studied how rural free delivery rolled out in America, and found that when it arrived in a new county, investment in manufacturing soon followed.[31] The same process seems to be unfolding in China, which has its 'Taobao Villages': clusters of rural enterprises producing everything from red dates to silver handicrafts to children's bicycles.[32]

Taobao is an online marketplace, owned by Alibaba. It's basically just a list of goods and prices. But perhaps it can aspire to shape society as much as any titan of literature – like Montgomery Ward.

14

Fast-Food Franchises

The way Ray Kroc tells the story, when he encouraged the McDonald brothers to open more hamburger restaurants, they winced.

The year was 1954. The place: San Bernardino, California, then a quiet town on the edge of the desert, some 50 miles east of Los Angeles. Kroc sold milkshake machines; Dick and Mac McDonald were among his best customers. Their restaurant was small, but it sold lots of milkshakes. Clearly, they were doing something right.

But they didn't want to do more of it, and Mac McDonald explained why: 'We sit out on the porch in the evenings and watch the sunset ... It's peaceful.'[1] Opening more branches would be a headache – travelling around, finding locations, vetting managers, staying in motels. Why bother? They were already making more money than they could spend.[2]

That might sound reasonable to many people, but not to Ray Kroc. 'His approach was utterly foreign to my thinking,' Kroc later recalled. He convinced the brothers to let *him* expand their restaurant chain. By the time he died, three decades later, McDonald's had thousands of restaurants bringing in billions of dollars.[3]

Which goes to show: successful entrepreneurs aren't all the same. They want different things. They have different talents.

Take Dick and Mac. They were brilliant at figuring out more efficient ways to make hamburgers. Working with a local craftsman, they invented a new kind of spatula; a new dispenser that squirted the same amount of ketchup and mustard every time; a rotating platform to speed up assembling of burger, bun and condiments. What Henry Ford had done for cars, the McDonald brothers did for hamburgers and French fries: they broke down processes into simple, repetitive tasks. This meant they could churn out food quickly, cheaply and consistently. There was nothing else like it.[4]

But when it came to the wider world, the brothers seem to have been fairly clueless. When competitors started peering through the windows, taking out notepads and sketching plans, Dick and Mac laughed about it.[5] When anyone asked about those ingenious condiment dispensers, they cheerfully named their craftsman friend. None of them had bothered to patent the design.[6]

Some people wanted more than snatched sketches, so the brothers sold franchises – after a fashion. For a one-off fee, you could buy blueprints to their building, with the golden arches, a 15-page description of their 'Speedy Service System', and a week's training. After that, the franchisees were on their own.[7]

Dick and Mac didn't expect that their trainees would serve the same menu, or even use the same name: when their very first franchisee mentioned that he would also call his new restaurant 'McDonald's', Dick replied, 'What the hell for?'[8]

Into this smooth-running kitchen and half-baked franchising operation walked a man with different skills and desires. Ray Kroc was in his fifties, and managing health problems from diabetes to arthritis.[9] But he was keener on money than peaceful sunsets, and he loved life on the road: Kroc later wrote, 'Finding locations for McDonald's is the most creatively fulfilling thing I can imagine.'[10] Where the brothers had rethought French fries, Kroc now rethought the franchise.

The idea itself wasn't new. The word 'franchise' comes from the old French 'franche', meaning 'free' or 'exempt'. In ages past, a monarch might grant you a franchise to organise a market, say – the exclusive right to do a certain thing in a defined area for a set time. In the nineteenth century, you might buy the exclusive right to sell Singer sewing machines in your local area.[11]

Nowadays franchise operations are everywhere. Stay in a Hilton or Marriott, rent a car from Hertz or Europcar, shop at a 7-Eleven or Carrefour, and you're likely to be dealing with a franchise owner.[12] This idea of the business-*format* franchise seems to have started in the 1890s with the Canadian Martha Matilda Harper, who built an international network of beauty salons: she was a former servant and her franchises transformed the lives of other servant girls.[13]

But it was 1950s fast food that gave the franchise its modern form, with not only McDonald's but Burger King, Kentucky Fried Chicken and many now-forgotten brands.[14] Ray Kroc's big insight was the importance of conformity.[15] You don't just sell the right to use the company's name and learn its methods – you impose an obligation to do things in a certain way. McDonald's opened a full-time training centre, 'Hamburger University', drilling students in subjects such as which kind of potatoes to buy.[16] Inspectors went round to write 27-page reports on whether franchisees cooked food at the right temperatures and kept the bathrooms clean.[17]

At first glance, the appeal to the budding restaurateur isn't obvious: wouldn't you want to design your own branding and develop your own menu? Why pay the McDonald's corporation $45,000, plus 4 per cent of gross sales,[18] just so they can send an inspector to watch you scrub your own toilets? Well, much of what you're paying for is the benefit of the brand – and if you're being monitored to make sure you don't cut corners that damage the brand, you can feel reassured that your fellow franchisees are, too.

As for the franchisor, why not own and operate new branches yourself? Many companies do both – McDonald's owns about 15 per cent of its 36,000-odd outlets.[19] But franchisees bring a lot to the company. There's cash – a McDonald's restaurant can cost more than a million dollars to launch.[20]

Franchisees also provide local knowledge, especially important if you're expanding into a new country with an unfamiliar culture. And there's motivation: an owner–manager with their own money at stake might put more effort into keeping costs down than a manager on a corporate salary. The economist Alan Krueger found evidence that may support this idea: workers and shift supervisors apparently earn more in company-owned fast-food outlets than in franchised ones.[21]

Of course, both sides bear some risk. The franchisor has to trust that the franchisee will work hard; the franchisee has to trust that the franchisor will create and advertise exciting new products. When both sides worry about the other side shirking, it's known as 'double-sided moral hazard'. A branch of economics called agency theory tries to understand how franchise contracts solve this problem through their mix of upfront fees and percentage payments.[22]

But it seems to work, perhaps because – like Kroc and the

McDonald brothers – different entrepreneurs want different things. Some people want the freedom to run their own business, day to day, but aren't interested in developing products or building a brand.

One of the McDonald brothers' early franchisees decided he didn't much like their golden arches, so he got his builders to make them pointy and named his restaurant 'Peaks' instead.[23] Those were freewheeling times; these days, the division of entrepreneurial labour is as regimented as a carousel full of hamburgers.

15

Fundraising Appeals

'It is not from the benevolence of the butcher, the brewer, or the baker that we expect our dinner,' wrote Adam Smith, famously, in *The Wealth of Nations*, 'but from their regard to their own interest. We address ourselves, not to their humanity but to their self-love, and never talk to them of our own necessities but of their advantages.'[1]

But when Smith was writing in the 1770s, his mail probably didn't include envelopes with arresting images of hungry children. When he strolled around his home town of Kirkcaldy, he was not accosted by clipboard-wielding young women trying to sign him up for a monthly donation. These days we are frequently spoken to not of our advantages, but of other people's necessities.

Charity has become big business, though it's hard to say how big: there's little good data. One recent study estimates that the British, for example, donate 54 pence in every £100; that's three times more than the Germans, but the Americans give three times more again.[2]

That's also roughly what Britons spend on beer; not much less than they spend on meat; and three times what they spend on bread.[3] In economic significance, the charity fundraiser is up there with the butcher, brewer and baker.

Charity, of course, is as old as humanity. The ancient religious custom of tithing – indirectly giving a tenth of one's income to worthy causes – makes modern donations of less than a pound in every hundred seem derisory.[4] Still, taxes have replaced tithes, and modern fundraisers don't have the advantage of claiming to speak for God. They need to be professional about persuasion, and there is a man who's regarded as the father of the field.[5]

His name is Charles Sumner Ward, and in the late nineteenth century he started work for the YMCA, the Young Men's Christian Association. The *New York Post* described Ward as 'a medium-sized man, so mild of manner that one would never suspect him of the power to sway hitherto reluctant pocketbooks'.[6]

That power first gained wide attention in 1905, when his employers sent him to Washington, DC, to raise money for a new building. Ward found a wealthy donor to pledge a chunk of cash, but only if the public raised the rest; then he set an artificial deadline for this to happen. The papers lapped it up: 'The YMCA's Fight Against Time to Raise $50,000', said one.[7]

Ward applied his methods far and wide: a target; a time limit; a campaign clock, showing progress; publicity stunts planned with military precision. In the modern world, they all seem familiar, but when Ward came to London in 1912, they were novel. *The Times* was suitably impressed by his 'knowledge of human nature and an extremely shrewd application of business principles in securing the advantage at the psychological moment'.[8]

The First World War brought more fundraising innovations: lotteries; and flag days, which have modern equivalents in wristbands, ribbons and stickers that show you've given money.[9] By 1924, Ward had a fundraising firm and was advertising how much it had raised for everyone from Boy Scouts to Masonic Temples; 'Campaigns conducted on a moderate fee for service rendered'.[10]

For the modern heirs of Charles Sumner Ward, what counts as a 'shrewd application of business principles'? We can get some clues from advertising executives interviewed for the *Guardian*. Images of starving children don't rack up many likes on social media, they say. Build your brand instead. Engage and entertain.[11]

Economists have also studied what motivates donations. One theory is called 'signalling': we donate in part to impress other people.[12] That would explain the enduring popularity of wristbands, ribbons and stickers: they display not only the causes that matter to us, but our generosity too.[13]

Then there's the 'warm glow' theory, which says that we give in order to feel nice – or less guilty, at least. Neither theory, you may notice, discusses whether or not the charities actually work.

Experimental investigations of these ideas have produced results that are – well, a little depressing. The economist John List and colleagues sent people to knock on doors; some asked for a donation, others sold lottery tickets for the same good cause. The lottery tickets raised a lot more; no surprise there.

But the researchers also found that attractive young women who asked for donations fared much better – about as well as the lottery sellers. As the study notes, dryly, 'This result is largely driven by increased participation rates among households where a male answered the door.'[14]

That's evidence for the signalling theory of altruism – and you can see exactly what kind of pretty young lady these gentlemen were keen to signal to.

Another economist, James Andreoni, studied the 'warm glow' idea by asking what happens to private donations when a charity starts getting a government subsidy. If donors gave purely from an altruistic desire to ensure the charity can function, then the donations should move to another worthy cause when the subsidy arrives. That doesn't happen, which suggests we aren't purely altruistic – we just get a warm glow from feeling that we are.[15]

It's starting to sound like Adam Smith's logic applies to charity after all. 'It is not from the benevolence of the donor that we expect a contribution,' a fundraiser might say, 'but from their regard to making themselves feel good or look good to others.'

But if charities are selling a warm glow and the ability to send social signals, that doesn't give them much incentive to do anything useful. They just have to tell us a good story.

Some people, of course, take very seriously the question of how much good charities do. There's a movement calling for 'effective altruism',[16] featuring organisations such as GiveWell, which studies charities' effectiveness and recommends who deserves our cash.[17]

The economists Dean Karlan and Daniel Wood wondered whether evidence of effectiveness would improve fundraising. They worked with a charity to find out. Some supporters got a typical mailshot: an emotional story about an individual beneficiary, Sebastiana: 'She's known nothing but abject poverty her entire life ... ' Others got the same story, with an additional paragraph noting that 'rigorous scientific methodologies' attest to the charity's impact.

The results? Some people who'd previously given big donations seemed impressed, and gave more. But that was cancelled out by small donors giving *less*.[18] Merely mentioning science seems to have punctured the emotional appeal, and lessened the warm glow.

Which may explain why GiveWell haven't even tried to assess the household names of the charity world – the likes of Oxfam, Save The Children and World Vision. In an exasperated blog post, they explain that such charities 'tend to publish a great deal of web content aimed at fundraising, but very little of interest for impact-oriented donors'.[19]

Or: 'never talk to them of our own effectiveness', as Adam Smith might have said.

16

Santa Claus

A curious ritual takes place each year in Japan. It's
called 'Kurisumasu ni wa kentakkii' – 'Kentucky for
Christmas' – the habit of eating Kentucky Fried Chicken on
24 December. The affair began as an inspired bit of marketing, when KFC noticed in the 1970s that expatriates who
craved Christmas turkey were turning to fried chicken as
the closest available substitute. Now it has become a popular
Japanese tradition; there are queues around the block, and
customers will pre-order their chicken as early as October.[1]

Christmas, of course, is not a religious holiday in Japan,
where only a tiny minority of the population is Christian. But
'Kurisumasu ni wa kentakkii' demonstrates how easily commercial interests can hijack religious festivals – from Diwali
in India to Passover and Rosh Hashanah in Israel, but most
notoriously, Christmas in America.

Why, after all, does Santa Claus wear red and white?
Many people will tell you that the modern Santa is dressed
to match the red and white colours of a can of Coke, and
was popularised by Coca-Cola's advertising in the 1930s.[2]
A good story, but the red-and-white Santa himself wasn't

created to advertise Coca-Cola – why, he was touting the rival beverage White Rock back in 1923.[3] Rudolph the Red-Nosed Reindeer was the one who was invented as a marketing gimmick.[4]

The modern Santa Claus is actually a century older, adapted from Dutch traditions in the once-Dutch city of New York, by prosperous Manhattanites such as Washington Irving and Clement Clarke Moore in the early 1800s. Irving and Moore also wanted to turn Christmas Eve from a raucous partying of street gangs into a hushed family affair, everyone tucked up in bed and not a creature stirring – not even a mouse.[5]

Moore – who penned the line "Twas The Night Before Christmas' in 1823 – did as much as anyone to create the American idea of Santa Claus, the patron saint of giving presents to everyone whether they want them or not. It was in the 1820s, too, that advertisements for Christmas presents became common in the United States; by the 1840s Santa himself was a frequent commercial icon in advertisements.[6] Retailers, after all, had to find some way to clear their end-of-year stock.

The gift-giving tradition took firm hold. Ten thousand people paid to see Charles Dickens give readings of his *A Christmas Carol* (a story light on biblical details and heavy on the idea of generosity) in Boston in 1867.[7] Down the coast in New York the same year, Macy's department store decided that it was worth keeping the doors open until midnight on Christmas Eve, for last-minute Christmas shoppers.[8] The next year, Louisa May Alcott's novel *Little Women* was published. Its first line: 'Christmas won't be Christmas without any presents.'

The Christmas blow-out, then, is not new. Joel Waldfogel, an economist and author of *Scroogenomics*, has been able to track the impact of Santa on the US economy back across the decades. By comparing retail sales in December with sales in November and January, Professor Waldfogel has estimated the size of the Christmas spending bump all the way back to 1935, the era of the Coca-Cola Santa. It may surprise some to know that relative to the size of the economy, Christmas spending was three times bigger then than now. What is an everyday indulgence today would have been a once-a-year treat back in the 1930s.[9]

Waldfogel has also compared the Christmas boom in the United States to other high-income countries around the world. Again – perhaps surprisingly – the December spending boom in the US is not particularly large relative to other countries. Portugal, Italy, South Africa, Mexico and the United Kingdom have the largest Christmas retail boom relative to the size of their economies; the US is an also-ran.[10]

In the grand scheme of things, Christmas is a modest affair, financially speaking. In the US, for every thousand dollars spent across the year, just $3 are specifically attributable to Christmas. After all, you were going to have lunch anyway, and pay your rent, and fill your car with petrol, and buy clothes to wear. For certain retail sectors, however – notably jewellery, department stores, electronics and useless tat – Christmas is a very big deal indeed. A small fraction of a big number is still a big number – Waldfogel reckons that at least $60 or $70 billion is spent on Christmas in the United States alone, and perhaps $200 billion around the world.

Is that money well spent?

'There are worlds of money wasted, at this time of year, in getting things that nobody wants, and nobody cares for after they are got.'[11] That was the author of *Uncle Tom's Cabin*, Harriet Beecher Stowe, back in 1850 – an early example of what is an annual complaint.

Economists and religious moralisers do not often find

themselves having common cause, but on the subject of Christmas we do: we think that a lot of Christmas spending is wasteful. Time, energy and natural resources are poured into creating Christmas gifts that the recipients often do not much like.

Santa's gifts rarely miss the mark; he is, after all, the world's number one toy expert.[12] The same cannot be said of the rest of us. Professor Waldfogel's most famous academic paper is 'The Deadweight Loss of Christmas', which simply tried to measure the gap between how much various Christmas gifts had cost, and how much the recipients valued them – leaving aside the warm glow of 'It's the thought that counts'. He concluded that the typical $100 gift was valued at – on average – just $82 by the recipient.[13]

This wastage figure seems to be fairly robust across countries – there is even a research paper by two Indian economists estimating the deadweight loss of Diwali[14] – and it suggests that $35 billion is being wasted around the world in the form of poorly chosen Christmas gifts. To put it into context, that is about what the World Bank lends to developing country governments each year.[15]

This is real money, and it's really being wasted. And that is before pondering the strain put on the economy by squeezing the retail spending together into a single month rather than spreading it out – and the time and aggravation devoted to the process of shopping, which is not always pleasant during the December rush.

So other economists have examined alternatives to clumsy gift-giving. Gift cards and vouchers do spare the waste of physical resources that an unwanted gift entails, but otherwise they do not help as much as one might hope: they are often unredeemed, or resold online at a discount. If you must buy a gift card, note that vouchers for lingerie sell well below face value on eBay, but vouchers for office supplies and coffee hold up pretty well.[16]

Wish lists fare better. Research suggests that recipients are generally delighted to receive a gift they have already specified; givers are deceiving themselves to think an off-piste gift will be more welcome.[17] Even Santa Claus likes to receive a polite wish list from good children. Who are the rest of us to think we can do better?

Or we could learn from the reformed Ebenezer Scrooge, who, Dickens declares, 'knew how to keep Christmas well, if any man alive possessed the knowledge'. On Christmas morning the only physical gift he gave was a prize turkey. The Christmas spirits had shown him that the turkey was sorely needed.

Other than that, he gave people his company and his money – including a rise for Bob Cratchit. Money! That's the true spirit of Christmas. God Bless us, every one!

III

MOVING MONEY

17

SWIFT

We don't always notice vital infrastructure until something goes wrong. So it was at Citibank London in the 1960s. On the first floor, payment instructions were inserted into a canister and sent upstairs via vacuum tube. On the second floor, a team of people confirmed the transactions and sent their authorisations back down the pipe.

One day the payments department on the first floor were receiving none of the authorisations they needed. Someone was sent upstairs, where the confirmation team had been idly wondering why all was quiet. It turned out that the vacuum tube had become blocked. Citibank's payment processing operations were duly restored with the assistance of a chimney sweep.[1]

Confirming large financial transactions is hard. It is even harder across national or international borders. Since the development of the telegraph in the first half of the nineteenth century, sending instructions has been quick enough, but quick does not necessarily mean foolproof, as Frank Primrose, a Philadelphia wool broker, was to realise.

In June 1887 Mr Primrose sent a message to his agent in

Kansas about buying wool. Because the Western Union telegraph company charged by the word, the message was in code
to save money. BAY ALL KINDS QUO, it was supposed to read.

It actually read BUY ALL KINDS QUO, so instead of understanding 'I've bought half a million pounds of wool', the
Kansas agent understood 'Please buy half a million pounds
of wool'. Primrose lost $20,000 – several million dollars in
today's terms. Nor would Western Union compensate him,
because he could have paid a few cents extra for the message
to be verified – but had not.[2]

Clearly, there was a need for a way to send financial information more reliably than a vacuum tube and more securely
than a telegraph in a code too easily mistranscribed.

For decades after the Second World War, banks used
telex machines, which made efficient use of telegraph lines
and allowed users to type a message somewhere and have it
printed on the other side of the world.[3] But the need to make
sure that messages were secure and accurate added enormous
complexity. Banks hired former military signalmen to operate their telex machines, and used tables of cross-referenced
codes to check and recheck what was being sent. One veteran
recalled the laborious complexities:

> For every single telex that was sent you had to manu-
> ally calculate what this telex test key was ... When you
> received the tested telex you had to do the reverse calcula-
> tion to make sure that the telex hadn't been tampered with
> during transmit and receive cycles ... It was incredibly
> prone to human error.[4]

By the globalising 1970s, the telex system was groaning
under the strain. That wasn't a problem for just banks, but also
for the rest of us. Products could be cheaper, better and more

varied if international trade could be supported by highly efficient bank transactions. The idea of travelling the world armed only with credit and debit cards – something we can take for granted today – depends on much better communication between banks than was feasible fifty years ago.

Especially in Europe, the need for a better solution – one that could work smoothly across borders – was becoming acute. Committees were established, arguments raged, progress was glacial. Then an American bank started strong-arming everyone into using its own proprietary system, called MARTI. One European banker recalled the tone of the US bank's demands:

> If you don't use it, we will not execute your instructions. If your instructions ... come via telex, we will return the telex. If we receive them by mail, we will put them into an envelope and send it back to you.[5]

This was, as they say in Europe, *insupportable*. Many banks feared becoming locked in to any standard that was owned by a rival. So they got their act together through a new organisation, SWIFT – the Society for Worldwide Interbank Telecommunication. SWIFT was a private company, headquartered in Brussels and run as a global cooperative venture, initially between 270 banks across 15 countries. The first SWIFT message was sent by Prince Albert of Belgium on 9 May 1977 – and the MARTI system closed down the same year.[6]

SWIFT simply provided a messaging service, using a standardised format that minimised errors and dramatically simplified proceedings. The computing company Burroughs installed SWIFT's dedicated computers and connections in Montreal, New York and 13 European banking centres. Each nation's banks would plug into those central hubs.[7]

The underlying hardware and software continues to change, transmitting and storing over six billion highly sensitive cross-border banking instructions a year. More important than any particular technology is the cooperative structure of the institution, in which now 9000 member banks and other institutions agree standards and resolve disagreements.[8]

Hacks, outages and other problems have occurred – often the result of weaknesses in the systems of banks from smaller or poorer nations.[9] Yet they remain rare enough for SWIFT to seem indispensable. The organisation itself would prefer to stay under the radar, a humble part of the financial pipework, operating out of a lakeside office in the sleepy town of La Hulpe, near Brussels.[10]

But having largely solved one problem, SWIFT may have created another. It is so central to international banking that it is a tempting tool for the 800lb gorilla of the world economy, the US government. Want to track terrorist financing? Examine the SWIFT database.[11] Want to destroy the Iranian economy? Tell SWIFT to deny access to their banks. After all, as any London chimney sweep can tell you – financial pipework can be blocked.

SWIFT has found itself unable to resist direct orders from the US, even when the EU disagrees.[12] The US has this power because a transaction between – say – a German lens maker and a Japanese camera maker will be converted from euros and yen into dollars, the universal medium of commerce. The messaging system is organised out of Brussels, but the deal will be settled in the US by US banks or the US subsidiaries of international banks. The United States government thus gets to see a great deal of information and to sanction any bank that displeases it.[13] SWIFT isn't interested in geopolitics, but geopolitics is interested in SWIFT.

The political scientists Henry Farrell and Abraham

Newman see the argument over SWIFT as an example of what they call 'weaponised interdependence' – the big boys of the global economy using their influence over supply chains, financial settlement and communications networks to monitor and punish wherever they wish. The US's blacklisting of the Chinese telecoms firm Huawei is another example.[14]

This isn't a completely modern tactic. In 1907, after a severe banking crisis had rocked the US and left the British financial system largely intact, British strategists took note. The UK was losing ground as a manufacturing economy, but as a financial hub it remained supreme. The City of London sat at the centre of a web of banks, telegraph lines and the deepest insurance market in the world. The thinking was that in a war, Germany's banks could swiftly be crushed by financial shock and awe.[15]

Spoiler alert: the plan did not work. But that historical parallel is unlikely to frighten the US. It is likely to keep a firm grip on the pressure points of the international economy – including the SWIFT messaging system. For an organisation that was galvanised as a response to pushy Americans, that is quite a kink in the financial pipe.

18

Credit Cards

The clue is in the name: credit. It means belief; trust. And the story of the modern economy couldn't be told without a chapter on who we trust, and how we come to trust them.

That was once an easy enough question to answer: trust was personal, a bond between two people who knew each other and had faith that a debt would be repaid. These days, trust takes another form: a stiff plastic rectangle with rounded edges, 3⅜ inches long, 2⅛ inches wide, and ¹⁄₃₂ of an inch thick. A credit card.

But I'm getting ahead of myself. In the days before trust was something you could slip into a slim wallet, people would get credit from a community store – from a shopkeeper who knew them, and where they lived, and that if they didn't pay their debts, the shopkeeper could complain to their mothers about them in church on Sunday.

As cities boomed in the early twentieth century, things became awkward. A large department store might be happy to extend credit, but shop assistants couldn't recognise each customer by sight. And so retailers would issue tokens – coins, key-rings, even objects resembling dog-tags called 'charga-plates'.[1]

With hindsight, this was a significant step: credit was being depersonalised so that a shop assistant could allow a person he or she did not recognise to walk out of the store with an armful of goods not yet paid for. And perhaps revealingly, some of the credit tokens became status symbols. They signified, 'I am the kind of person that is trusted'.

But the technology of trust could be broadened further, by introducing a charge token that allowed someone to get credit not just from a single store but from a range of stores. The first example was Charg-It; this universal token appeared in Brooklyn, also in 1947 – although it was only universal within a two-block area.

Hard on its heels came Diners Club, founded in 1949. The founding myth of the Diners Club card is that a businessman named Frank X. McNamara found himself embarrassed after taking clients to dinner and realising he'd left his wallet in another suit. The tale may well be fictional, but regardless, McNamara conceived of a card that would be an essential tool in the pocket of a travelling salesman, allowing him to buy food and fuel, rent hotel rooms and entertain clients. It would work not just at a single department store, but at a network of outlets around the United States. The Diners Club card took off, with 35,000 subscribers in the first year. The company rushed to sign up hotels, airlines, petrol stations and car hire firms – as well as expanding into Europe.[2]

The Diners Club card wasn't yet a credit card. It was a charge card; the balance had to be paid off promptly and in full every month; credit was extended as a side-effect of making it easy to run a corporate expense account.

But the true credit card wasn't far behind. By the end of the 1950s Diners Club was competing with the courier and provider of traveller's cheques, American Express, and with credit cards set up by the banks. Most prominent was Bank of America, with its BankAmericard. BankAmericard would eventually become Visa; its rival, Master Charge, became MasterCard. These cards added a rotating credit balance: you didn't have to pay off your debts in full; you could roll them over.

Credit cards had to overcome a chicken-and-egg problem: retailers didn't want to go to the trouble of accepting them unless lots of customers demanded it, while customers couldn't be bothered signing up for the cards unless plenty of retailers would accept them.

To overcome the inertia, in 1958 Bank of America took the bold step of simply mailing a plastic credit card to every single Bank of America customer in Fresno, California – 60,000 of them. Each card had a credit limit of $500, no questions asked – closer to $5000 in today's terms. The audacious move was known as the Fresno Drop, and it was quickly emulated, despite the obvious – and expected – losses from delinquent loans and outright fraud by criminals who simply stole the cards out of people's mailboxes.[3] The banks swallowed the losses, and by the end of 1960, Bank of America alone had a million credit cards in circulation.[4]

A cultural shift was well underway: despite the existence of prestige products such as Platinum cards, the credit card was no longer the preserve of a financial elite. It was an everyday thing, marketed to students and to divorcees to smooth over temporary financial embarrassment. Anyone could have one; anyone could be trusted. The credit card didn't require you to genuflect to a bank manager as you begged for a loan and explained yourself. You could spend on anything you wanted, and repay the debt at your own convenience – as long as you didn't mind paying interest rates that could easily be 20 or 30 per cent.

But they were still a hassle to use: pull out a credit card

and the shop assistant would have to phone up your bank to get the transaction approved. New technologies helped to make the process of spending ever more painless. One of them was the magnetic stripe – originally developed in the early 1960s by Forrest and Dorothea Parry for use on CIA identity cards. Forrest was an IBM engineer who came home one evening with a plastic card and a strip of magnetic tape, trying to figure out how to attach one to the other. His wife Dorothea, who was ironing at the time, handed him the iron and told him to try it. The combination of heat and pressure worked perfectly and the magnetic stripe was born.[5]

Thanks to the stripe you could swipe a Visa card in a shop; the shop would send a message to its bank; the shop's bank would send a message to the Visa network computers; the Visa computers would send a message to *your* bank. If your bank was happy to trust you to repay, nobody else had to worry: the digital thumbs-up would pass all the way back through these computers to the shop, which would issue a receipt and let you walk out of the door. The whole process would have taken just a few seconds.[6]

With a contactless card, the process became even faster – faster than cash, which is becoming an obsolete technology in some countries. In Sweden only 20 per cent of payments at shops are made with cash, and across the economy, just 1 per cent by value.[7] Back in 1970, a BankAmericard advertising slogan had been 'think of it as money'.[8] Now, for many transactions, physical money won't do: an airline or a car hire firm or a hotel want your credit card, not your cash – and in Sweden the same is true even of coffee shops, bars and sometimes market stalls.

So now the credit card is everywhere – and anyone who uses the technology can tap into a network of trust that was once the preserve of upstanding members of a tight-knit community. We can all enjoy the benefits of being trusted.

But having such effortless, impersonal credit on tap might be doing strange things to our psychology. A few years ago, two researchers from MIT, Drazen Prelec and Duncan Simester, ran an experiment to test whether credit cards made us more relaxed about spending money. They allowed two groups of subjects to bid in an auction to buy tickets for popular sports fixtures. These tickets were valuable, but exactly how valuable wasn't clear. One group were told that they had to pay with cash – but not to worry, there was an ATM around the corner if they won. The other group were told that only payment by credit card would be accepted. There was a striking difference in the results: the credit-card group bid substantially more for the tickets, more than twice as much in the case of a particularly popular match.[9]

Credit cards can, used wisely, help us manage our money. The risk is that the credit card makes it simply too easy to spend money – money that we don't necessarily have. Rotating credit – that distinctive feature of a credit card – is now around $860bn in the United States, more than $2500 for every adult in the country. In real terms it has expanded four-hundred-fold since 1968.[10] And a recent study by the International Monetary Fund concluded that household debt – the kind of debt credit cards make it easy to accumulate – was the economic equivalent of a sugar rush. It was good for growth in the short term, but bad over a three- to five-year horizon – as well as making banking crises more likely.[11]

If you ask people about all this, they worry. Faced with the statement 'credit card companies make too much credit available to most people', nine out of ten Americans with credit cards agree; most of them strongly agree. Yet when they reflect on their own cards, they're satisfied.[12]

We don't trust each other to wield these powerful financial tools responsibly, it seems. But we do trust ourselves. I wonder if we should.

19

Stock Options

The average CEO at a major American corporation, according to a recent Senate hearing, is paid about a hundred times as much as the average worker [...] and our government today rewards that excess with a tax break for executive pay, no matter how high it is. That's wrong. If companies want to overpay their executives and underinvest in their future, that's their business, but they shouldn't get any special treatment from Uncle Sam.[1]

That was Bill Clinton, campaigning to be US president, in 1991. He won, of course. And he promptly made good on his promise to tackle excessive pay.

Usually, the salaries a company pays are treated as costs, reducing the profit on which it pays tax. Clinton changed the law: companies could still pay as much as they wanted to – but salaries over a million dollars would no longer be tax-deductible.[2]

It had a big impact. By the time Clinton left office, in 2000, the ratio of CEO pay to worker pay was no longer a hundred to one. No: it was ... well over three hundred to one.[3]

What had gone wrong? We can approach that question from the olive groves of ancient Greece.

The philosopher Thales of Miletus, so the story goes, was being challenged to prove the value of philosophy: if it was so useful, why was Thales so poor? Aristotle, who tells this story,[4] makes clear that the question is gauche: of course philosophers are smart enough to get rich; but they're also wise enough not to bother. We can imagine Thales heaving a sigh: okay, I'll make a fortune – if I must.

Philosophy, back then, included reading the future in the stars. Thales foresaw a bumper harvest of olives. That would mean high demand to rent time on the town's olive presses. Thales visited each press owner with a proposition. Aristotle is hazy on the details, but mentions the word 'deposit' – perhaps Thales negotiated the right to use the press at harvest time; and, if he decided against, the owner would simply keep his deposit.

If so, it's the first recorded example of what we now call the option.[5] A poor olive harvest, and Thales' option would be worthless. But, whether by luck or astronomical judgement, he was right. Aristotle tells us that Thales hired out the presses 'on what terms he pleased and collected a good deal of money'.[6]

The idea of the option pops up through history, from the Medicis in Florence to the Dutch tulip episode.[7] Nowadays, many options are bought and sold on the financial markets.[8] If I believe that Apple's share price will rise, I could simply buy Apple shares – or I could buy an option to buy Apple shares at a specified price on a future date.

The option is higher-risk and higher-reward. If the share price is lower than my option to buy, I've lost everything. If it's higher, I can exercise the option, resell the shares, and make a bigger profit.

But there's another use for stock options – an attempt to solve what economists call the principal–agent problem. A 'principal' owns something; they employ an 'agent' to manage it for them.

Imagine I'm made CEO of Apple, and you own Apple shares. That makes you the principal, or one of them. I'm the agent, managing the company for you and the other shareholders. You'd like to trust me to work hard in your interests, but you can't see what I'm doing all day. Maybe I make every decision by consulting an astrologer – not a smart one like Thales, either – but I always spin some plausible excuse for why profits are stagnating.

But what if I'm given options to buy new Apple stock in a few years? Now I stand to gain from making the share price rise. Sure, if I exercise my option, that slightly dilutes the worth of your shares – but if their price has been rising, you shouldn't mind.

It all sounds perfectly sensible, and in 1990 the economists Kevin J. Murphy and Michael Jensen published an influential paper on the topic. 'In most publicly held companies,' they wrote, 'the compensation of top executives is virtually independent of performance.' No wonder CEOs acted like 'bureaucrats' rather than 'value-maximizing entrepreneurs'.[9]

So when President Clinton cut tax breaks for executive pay, he exempted performance-related rewards. Clinton adviser Robert Reich, who opposed the exemption, explains what happened: 'It just shifted executive pay from salaries to stock options.'[10]

Over Clinton's term in office, the value of options granted to employees at top American companies grew tenfold.[11] A rising stock market meant even a horoscope-consulting CEO would have been quids in. The gap between bosses' and workers' pay ballooned. A Clinton-era congressman says the law 'deserves pride of place in the Museum of Unintended Consequences'.[12]

But hold on – if options incentivise executives to do a better job, surely that's no bad thing? Unfortunately, that turned out to be a big 'if'. One problem: what options really incentivise is maximising a company's share price on a given date. If you think that's exactly the same thing as running a company well, I have some shares in Enron to sell you.[13] Outright fraud aside, the stock options create temptation to be less than transparent about news that could weigh on the share price.[14]

If stock options aren't the best way to reward performance, shouldn't company boards of directors be keen to find alternatives? In theory, yes – it's the board's job to negotiate with CEOs on behalf of shareholders. In practice, this is another principal–agent problem, as CEOs can often influence who directors are and how much they're paid. There's obvious potential for mutual back-scratching.

In their book *Pay Without Performance*, Lucian Bebchuk and Jesse Fried argue that directors don't actually care about linking pay to performance, but must 'camouflage' this indifference from shareholders.[15] 'Stealth compensation' is the best form of compensation for fat cats, and stock options seem to be a way to achieve that.[16]

Perhaps shareholders need yet another agent to supervise how directors reward CEOs. There is one candidate: many people hold shares not directly, but through pension funds; there's some evidence that these institutional investors can persuade boards to be tougher negotiators.[17] When a large shareholder can assert some control, there's a more genuine link between executive pay and executive performance.[18] However, this link seems all too rare.[19]

Executive pay is often in the headlines, even in countries

where the gap to worker pay is less than in America.[20] Given this, there's surprisingly little evidence on what makes sense.[21] How well can you evaluate the job a CEO is doing? Opinions differ.[22] Were bosses in the 1960s really less motivated because they earned a mere 20 times workers' pay?[23] It seems most unlikely. On the other hand, good decisions at the helm of a large company are worth a lot more than bad ones. So maybe those CEOs really are worth eight-figure compensation packages. Maybe.

But if so, that isn't clear to voters or workers, many of whom still feel the anger about 'excess' that President Clinton once voiced.[24] Perhaps CEOs should learn from Thales, who was clever enough to make more money, but wise enough to wonder if he should.

20

The Vickrey Turnstile

In the 1950s, the New York subway faced a problem that will be familiar to users of public transport all over the world.[1] At peak times it was overcrowded; at other times the trains were empty. The mayor commissioned a report, which concluded that the problem was that subway riders paid a flat fare. No matter where you boarded, how far you travelled, or when you made your trip, it would cost you ten cents.[2]

Might there be some more sophisticated approach? Perhaps so. A foreword to the report singled out one of the 17 authors:

> It is to such questions that Mr. Vickrey has addressed himself, and with a degree of skill which we predict will command the admiration of the reader. The abandonment of the flat-rate fare in favor of a fare structure which takes into account the length and location of the ride and the hour of the day is obviously a sensible step provided the mechanical problems involved can be solved.[3]

William Vickrey's basic idea was simple: when the trains were busy, charge more. When they were quiet, charge less.

The peaks in demand would become less spiky. The subway would be more comfortable and reliable, could carry more people without having to build new lines, and could raise more money, all at once. A great idea. But how to charge all these different prices? Not with an army of ticket clerks and inspectors; that would take too much time and money. Some automated solution must be found, and fortunately:

> Mr. Vickrey submits several highly interesting and stimulating suggestions – suggestions which in our judgment deserve the most careful examination and consideration.[4]

What was needed was a coin-operated turnstile that could charge different rates for different journeys at different times. But in 1952, this would not be simple.

To give a sense of the challenge, consider a dilemma faced by the Coca-Cola company. A Coke had cost a nickel – five cents – for decades. Coca-Cola would have liked to increase the price by a cent or two, but it couldn't. Why? Their 400,000 vending machines took only nickels, and redesigning them to take two different denominations of coin would be 'a logistical nightmare'. In 1953 Coca-Cola tried instead to persuade President Eisenhower, in all seriousness, to introduce a 7.5 cent coin.[5]

But Vickrey wasn't daunted, and he described a contraption to solve the problem:

> To have passengers put a quarter in the entrance turnstile, get a metal check with notches indicating the zone of origin, to be inserted in an exit turnstile which would, through electro-mechanical relays, deliver an appropriate number of nickels according to the origin and time of day.[6]

It sounds clever, and you may be wondering why you haven't heard of it. A clue comes from the title of a speech in which Vickrey gave that description: 'My Innovative Failures in Economics'. He began that speech thus:

> You are looking at an economist who has repeatedly failed in achieving his objective.

The variable-price electromechanical Vickrey Turnstile was never built. So why are you reading a chapter about a non-existent invention? It's because the idea itself was so important, even if it seemed too complicated to realise. Vickrey's fellow economists often said that he was just too far ahead of his time. He eventually won a Nobel memorial prize in 1996, just three days before his death.

Vickrey was proposing what is often called 'peak-load pricing' by economists, and 'dynamic pricing' by management consultants. In its simplest form it's an old idea. 'Early bird specials' – offering a cheap deal to restaurant diners at quiet times – date back to the 1920s. It's an easy sell to customers and requires no electromechanical wizardry.[7]

But the idea has appeal in far more complex settings. Whether you're running a subway system or an airline, trying to fill a concert hall or balance an electricity grid, it can be very costly to add extra capacity just to meet a short-term peak in demand – and it is wasteful to carry unused capacity around at other times. Varying prices makes sense.

US airlines were early pioneers, after being deregulated and forced to compete fiercely in the late 1970s. By 1984, the *Wall Street Journal* reported that Delta Air Lines alone employed 147 staff to incessantly tweak prices.[8]

'We don't have to know if a balloon race in Albuquerque or a rodeo in Lubbock is causing an increase in demand for a flight,' said Delta's pricing guru Robert Cross. They just had to adjust prices profitably, ensuring that their planes neither sold out too cheaply nor took off empty.[9]

Peak-load pricing no longer requires an army of pricing specialists. A company such as Uber can effortlessly match supply and demand with an algorithm. Uber's 'surge pricing' promises to end that painful three-hour wait for a taxi on New Year's Eve; there's always a price at which you can get a car right now.

But consumer acceptance may be more of a problem. 'You're almost at their mercy because you don't want to wait longer for a cab,' whined one passenger after paying $247.50 for a 21-kilometre ride in Houston, Texas – although he only had to pay because he couldn't bear to wait.[10]

Consumers feel exploited by some forms of dynamic pricing – especially when, as with Uber, prices could double or halve in a matter of minutes.[11] A 1986 study by the behavioural economists Daniel Kahneman, Jack Knetch and Richard Thaler showed that people found price surges infuriating. This was true even in scenarios where the logic was obvious, as with a higher price for snow shovels after a snowstorm.[12]

Having once despaired over the lack of a 7.5 cent coin, Coca-Cola pushed technology beyond what customers would swallow when, in 1999, it flirted with a vending machine that on sweltering days would raise the price of an ice-cold Coke.[13]

And perhaps we are right to be wary. Delta's Robert Cross later went on to publish a book about dynamic pricing with the subtitle 'Hardcore Tactics for Market Domination'.

Some businesses have eschewed peak pricing entirely – for example, Japan's reliable, profitable and privately owned railway companies do not distinguish between peak and off-peak fares, which goes some way to explaining why Tokyo's subway rush hour is so legendarily packed.[14]

But peak-load pricing is likely to play an increasing role in the economy of the future. Consider a smart electricity grid fed by intermittent power sources such as wind and solar power. When a cloud covers the sun, your laptop might decide to stop charging, your freezer switch itself off for a minute, and your electric car might even start pumping energy into the grid rather than sucking it out. But all that would require those devices to respond to second-by-second price changes.

A favourite example of William Vickrey's was congestion pricing on roads, designed – just as the turnstile was – to smooth out demand and ensure that limited capacity was used well. That's now becoming a practical idea – drivers near Washington, DC, can switch into a free-flowing lane if they're willing to pay the variable charge, which can be as much as $40 for 10 miles when traffic is particularly bad.[15]

Vickrey had tried to show that it could work in the mid-1960s: he built a prototype, using a simple computer and a radio-transmitter to tally every time he used his own driveway.[16] But sometimes good ideas just need to wait for the technology to catch up.

21

The Blockchain

The Long Island Iced Tea Company, as its name suggests, sold beverages. And not as many as it might have liked: it lost nearly $4 million in the third quarter of 2017. Then the company made a grand and rather cryptic announcement. It would henceforth be known as the Long Blockchain Corporation. Would it stop selling beverages? No. It would still do that. Would it sell beverages using blockchain? Well, maybe. It would do *something* to do with blockchain. Probably. The details were hazy. But that didn't stop investors getting excited. The company's share price almost quadrupled.[1]

In a book about things that made the modern economy, one might reasonably query whether blockchain yet merits the use of the past tense. But venture capitalists are pouring billions into start-ups with more plausible-sounding plans than the Long Island Iced Tea Company's.[2] And billions more are being raised in the regulatory grey area of initial coin offerings.[3] Enthusiasts say blockchain could become as disruptive as the internet. Indeed, blockchain is often compared to the World Wide Web in the 1990s: back then, it seemed clear that this Interweb technology would become important – but few really understood it, or foresaw its potential and limitations.

So let's try to get our heads around blockchain. We can start with a deceptively simple question: what stops me from spending the same money twice?

When money meant coins, that was easy – I can't give the same coin to two people. But we long ago realised that lugging coins around is no way to run an economy. It's easier to trust intermediaries to keep records of who's got what. You give me goods; I instruct the record keepers to shuffle their numbers accordingly. How do you know I haven't promised the same money to someone else? You trust the bank, or MasterCard, or PayPal, to guarantee that can't happen – because their systems won't allow it, or they've satisfied themselves that I'm not that sort of chap.

This all works well enough. But it has some drawbacks. These intermediaries need paying for their services. Network effects often give them market power. They know more about us than we do about each other; that's another source of power. And if they fail, the whole system collapses.

What if we didn't need them? What if the financial records that lubricate the economy could somehow be communally owned and maintained?

In 2008, someone using the pseudonym Satoshi Nakamoto proposed a new kind of money: Bitcoin.[4] Transactions would be verified not by a trusted intermediary, but by a network of computers solving cryptographic puzzles. If anyone ever controlled most of this network, they could fake the records and defraud people by double-spending Bitcoins – but that couldn't happen as long as enough different people chipped in computing power to check the solutions. And people would be incentivised to contribute computing power by random occasional rewards in Bitcoin.

It was ingenious. And people soon noticed that the underlying technology might have wider applications. It offered a completely new way for strangers to collaborate without needing to trust an intermediary or centralised authority. We started to hear phrases like 'transform everything' and 'change the world'.[5]

That underlying technology is known as blockchain, because blocks of transactions are periodically approved by the network and added to a public chain of records. It's also known as distributed ledger, because – well, it distributes the ledger: every participant keeps their own copy of those records. The economists Christian Catalini and Joshua Gans describe blockchain as a general-purpose technology that can lower the costs of verifying transactions and lower the barriers to creating new marketplaces.[6] In principle, blockchain might be useful in any situation where we currently trust some entity to manage our data in ways that help us to interact.

When you think about it, surprisingly many situations fit that description. What are Facebook, Uber and Amazon, for example, if not databases that help us interact? Might blockchain one day build new online models where we own our data, or perhaps sell our attention directly? Some think so.[7] Others are working on blockchains to track goods through supply chains, or intellectual property in the digital world; to make contracts quicker to administer, or voting systems more secure. You name it, somebody somewhere will be trying to put it on a blockchain.

But let's be honest: most of us don't understand the nuts and bolts of these ideas. And even if we do, we can't confidently envisage how they'll play out in reality. Predictably, the combination of intense buzz and hard-to-grasp technology has led some people not to think as critically as perhaps they might. The kind of people who'd rush to buy shares in a loss-making drinks company when it puts the word 'blockchain' in its name. Or sink $660 million into something called Pincoin, seemingly based on little more than a glitzy, buzzword-laden website; the people behind Pincoin appear to have taken the money and done a runner.[8]

Just how excited should blockchain make us? The economist Tyler Cowen is cautious: he reckons 'skepticism is more plausible than enthusiasm' – for now, at least.[9] One reason: blockchains can be slow and power-hungry. Bitcoin, for example, chugs through three or four transactions a second; by contrast, Visa averages 1600.[10] To validate these transactions, the computers solving Bitcoin's cryptographic puzzles consume, by one estimate, about as much electricity as Ireland.[11]

Some dispute the significance of these figures, but the technological challenge of scaling blockchains seems real.[12] So is the problem of marrying data to real-world stuff, or humans. It's one of Bitcoin's attractions that your wallet isn't linked to your real identity – especially if you're using it to purchase dodgy stuff. But if we want to use blockchains to store medical records, say, we have to be certain that they can't get attached to the wrong patient.[13]

In removing the need for intermediaries, blockchains may sometimes remind us why we find their services worth paying for.[14] Intermediaries can rectify mistakes: lose your internet banking passcode, and your bank will send you a new one; lose the passcode to your Bitcoin wallet, and you can kiss your Bitcoins goodbye.[15] Intermediaries can resolve disputes; how best to do that with blockchain 'smart contracts' is an evolving conversation.[16]

And trust in an intermediary has to be replaced by trust in other things – that software isn't buggy, and incentive structures won't break down in unexpected circumstances.

But auditing code is hard: the Decentralized Autonomous Organization, a pioneering investment fund on the Ethereum blockchain, raised $150 million – then someone hacked it, and nicked $50 million.[17] The economist Eric Budish has suggested there are limits to how valuable Bitcoin can become before the incentives to attack it outweigh the incentives that currently keep attackers at bay.[18]

But it's barely a decade since blockchain was invented. Shouldn't we expect some wrong moves and false starts before we figure out what it's good for? When the World Wide Web was a similar age, investors were pouring money into Webvan, Flooz and Pets.com, as well as eventual successes like Amazon.[19] It shouldn't surprise us that shares in Long Blockchain quickly crashed by 96 per cent.[20] But nor should it make us too cynical about what might one day be possible.

IV

INVISIBLE SYSTEMS

22

Interchangeable Parts

One sweltering afternoon in July 1785, officials, dignitaries and a few infuriated gunsmiths gathered at the
Château de Vincennes, a spectacular castle to the east of Paris.
They were there to see the demonstration of a new sort of
flintlock musket designed by Honoré Blanc, a gunsmith from
Avignon so despised by his fellows that he had been holed
away in the dungeons of the château for his own protection.[1]

Down in the cool of the castle cellars, Monsieur Blanc
brought in 50 locks – the lock being the firing mechanism
at the heart of a flintlock weapon. Briskly he took apart half
of them and, with the insouciance for which the French are
famous, he tossed their component parts into boxes. There
was a box for the mainsprings, a box for the hammers, a box
for the faceplates and a box for the gunpowder pans.[2]

Then, like a master of ceremonies ostentatiously agitating
an urn full of numbered lottery balls, Monsieur Blanc shook
these boxes to mix their components together. Then he
calmly pulled out the parts at random and began to reassem-
ble them into flintlocks.

What was he thinking? Everyone present knew that each

hand-crafted gun was unique. You couldn't just jam a part from one gun into another and expect either to work. But they did. Blanc had taken enormous pains to ensure that all the parts were precisely the same.[3]

It was a spectacular demonstration of the power of inter-changeable parts. The implications weren't lost on one of the visiting dignitaries: the emissary to France – and the future president – of the fledgling nation of the United States of America, Thomas Jefferson.[4]

Jefferson excitedly wrote to the US foreign secretary John Jay:

An improvement is made here in the construction of the musket which it may be interesting to Congress to know ... it consists in the making every part of them so exactly alike that what belongs to any one may be used for every one musket in the magazine ... I put several together myself taking pieces at hazard as they came to hand, and they fitted in the most perfect manner. The advantages of this, when arms need repair, are evident.[5]

But perhaps they were not evident, since Jefferson struggled to get his colleagues to embrace the idea. He wrote repeatedly to Henry Knox, the US Secretary of War in whose honour Fort Knox is named, trying to persuade him to hire Honoré Blanc and introduce his system. Knox did not respond.[6]

So what exactly were the 'evident' advantages of this system? Jefferson focused on the problem of battlefield repair. A cracked mainspring or warped gunpowder pan could render a soldier's gun useless. Fixing it would mean hand-crafting a new part to fit its brothers and sisters perfectly, a task requiring complex equipment and hours of skilled labour.

But under Blanc's system, only a few minutes and some

rudimentary skill would be required to unscrew the musket, replace the faulty part with an identical component, and screw it all back together, good as new. No wonder Blanc's fellow gunsmiths were worried about the future of their profession. And no wonder Thomas Jefferson was so interested in the problem of repairing broken guns.

But while Jefferson struggled to win support, Honoré Blanc was struggling too. It was impossibly expensive to hand-craft each piece to the precision required to make the system work.

The solution already existed, if only Blanc had grasped it. It would not only allow the swift repair of broken weapons, but a revolution in the world economy. A decade prior to Blanc's demonstration, a gentleman nicknamed John 'Iron-Mad' Wilkinson was becoming a local celebrity – 'local' being Shropshire, on the border between England and Wales – for his iron boat, iron pulpit, iron desk and even iron coffin, out of which he would burst to surprise visitors.[7]

He deserves far more fame for, in 1774, inventing a method of boring a hole into a cannon-shaped lump of iron so that it was straight and true every single time. That was militarily invaluable. But Iron-Mad Wilkinson hadn't finished. A few years later, he ordered one of these new-fangled steam engines from a neighbouring business. They were having trouble making it work, though: the piston cylinder, formed of hand-beaten panels of metal, didn't have a perfectly circular cross-section, and so steam leaked out everywhere around the piston head.[8]

Give it here, said John Wilkinson, and he used his cannon-boring method to make a pleasingly round piston cylinder.[9] His supplier, a Scotsman named James Watt, never looked back. Supplied with Watt's brilliantly efficient steam engines, running with Wilkinson's precisely bored cylinders, the Industrial Revolution entered a higher gear.[10]

Wilkinson and Watt weren't worried about interchangeable parts, as such. They wanted cannonballs to fit into cannons, and pistons to fit into cylinders. But the engineering problem they were solving also held the key to the interchangeability that Blanc prized but was finding it expensive to achieve. Wilkinson had built a machine tool – a tool that automates a manufacturing process – which comprised a very sharp drill, a water-mill, and a system of clamping one thing while smoothly rotating another.[11]

Swift on his heels was Henry Maudslay, at first a brilliant London locksmith's even more brilliant apprentice, and then the designer of unprecedentedly precise machine tools, designed to perform the same process again and again with exactitude. In the early 1800s they were taken up by the Royal Navy to make the pulley blocks that raised and lowered the sails of navy battleships.[12]

But as Simon Winchester observes in his history of precision engineering, *Exactly*, these machine tools had a curious side-effect: the Block Mills of Portsmouth produced the most perfect pulley blocks ever seen, but they also put skilled craftsmen out of work in large numbers. Honoré Blanc's fellow gunsmiths had been worried that they would lose out on lucrative repair work. But they were about to lose manufacturing jobs, too. Not only were machine tools better than hand tools, but they also did not require hands to wield them.

There was a second unlooked-for consequence. If you could use machine tools to produce perfectly precise interchangeable parts, that not only made for simple battlefield repair – as Jefferson had seen – but it also made the process of assembly simpler and more predictable. Adam Smith's famous description of a pin factory, nine years before Blanc's demonstration, depicted each worker adding a step to what had come before.[13] But with interchangeable parts, such a production line could become a far quicker, more predictable and more automated process.[14]

Across the Atlantic, the Americans had finally started to listen to Jefferson, and the promise of the system was eventually realised at an armoury at Harper's Ferry in Virginia, which in the 1820s began to produce, in Winchester's words, 'the first truly mechanically produced production line objects made anywhere'. As Honoré Blanc had always intended, they were guns, lock, stock and barrel.[15]

It was the beginning of what became known as the 'American system of manufacturing', which over the course of the next century came to produce Cyrus McCormick's reapers and harvesters, Isaac Singer's sewing machines and Henry Ford's Model T. Ford was a champion of interchangeability, and the Model T production line would have been inconceivable without precisely machined interchangeable parts.[16]

As for poor Honoré Blanc, he was undone by the French Revolution of 1789 – his dungeon workshop sacked by a mob, his political support guillotined. He struggled on, hopelessly in debt. Blanc had given birth to an economic revolution – but thanks to a revolution of a very different kind, he never saw his own ideas realised.[17]

23

RFID

It was 4 August 1945. The European chapter of the Second World War was over. The USA and the USSR pondered their future relationship. At the American embassy in Moscow, a group of boys from the Young Pioneer Organization of the Soviet Union made a charming gesture of friendship between two superpowers: they presented a large, hand-carved ceremonial seal of the United States of America to Averell Harriman, the US ambassador. It was later to become known simply as the *Thing*.[1]

Naturally, Harriman's office would have checked the heavy wooden ornament for bugs, but with neither wires nor batteries in evidence, what harm could it do? Harriman gave the *Thing* pride of place, hanging on the wall of his study – from where it betrayed his private conversations for the next seven years. He couldn't have realised that the device had been built by one of the true originals of the twentieth century.

Leon Theremin was famous even then for his eponymous musical instrument. He'd been living in the US with his African-American wife, Lavinia Williams, before returning to the Soviet Union in 1938 – kidnapped, she said. In any case, he was promptly put to work in a prison camp, where he was forced to design, among other listening devices, the *Thing*.[2]

Eventually, American radio operators stumbled upon the US Ambassador's conversations being broadcast over the airwaves. These broadcasts were unpredictable: scan the embassy for radio emissions, and no bug was in evidence. It took yet more time to discover the secret. The listening device was inside the *Thing* – and it was ingeniously simple, little more than an antenna attached to a cavity with a silver diaphragm over it, serving as a microphone. There were no batteries or any other source of power. The *Thing* didn't need them. It was activated by radio waves beamed at the US embassy by the Soviets, at which point it would broadcast back, using the energy of the incoming signal. Switch off that signal, and the *Thing* would go silent.

Much like Leon Theremin's unearthly musical instrument, the *Thing* might seem a technological curiosity. But the idea of a device that is powered by incoming radio waves, and which sends back information in response, is much more than that.

The RFID tag – short for Radio Frequency Identification – is ubiquitous in the modern economy. My passport has one. So does my credit card, enabling me to pay for small items simply by waving it near an RFID reader. Library books – and not only *RFID Essentials*, which I used to research this chapter – often have RFID tags. Airlines are increasingly using them to track luggage; retailers, to prevent shoplifting.[3] Some of them contain a power source, but most – like Theremin's *Thing* – are powered remotely by an incoming signal. That makes them cheap – and being cheap has always been a selling point.[4]

A form of RFID was used by Allied planes during the

Second World War: radar would illuminate the planes, and a substantial piece of kit called a transponder would react to the radar by beaming back a signal that meant 'we're on your side, don't shoot'. But as silicon circuits began to shrink, it became possible to conceive of a tag that you might attach to something much less valuable than an aeroplane.

Much like barcodes, RFID tags could be used to quickly identify an object. But unlike barcodes, they could be scanned automatically, without the need for line of sight. Some RFID tags could be read from several feet away; some could be scanned, albeit imperfectly, in batches. Some could be rewritten as well as read, or remotely disabled. And they could store much more data than a humble barcode, enabling an object to be identified not just as a particular type of comfort-fit size-medium jeans, but as a unique pair made in a certain place on a certain day.[5]

RFID tags were used to keep tabs on railway carriages and dairy cattle in the 1970s – the plastic tags were attached to cows' ears. By the 1980s, they were plotting the path of automobile chassis along the assembly line, the precursor to many 'closed-loop' RFID applications that track tools and materials throughout a complex production process.[6] Norway used RFID to automate road rolls at toll gates in 1987; by 1991, Oklahoma was using the technology to collect tolls without any need for cars to slow down.[7] In the early 2000s, large organisations such as Tesco, Wal-Mart and the US Department of Defense started demanding that their suppliers tag pallets of supplies – the endgame seemed to be an RFID tag on everything. A few enthusiasts even implanted RFID tags into their bodies – enabling them to unlock doors or ride the subway with a wave of the hand.[8]

In 1999, Kevin Ashton, an executive at fast-moving consumer-goods company Procter & Gamble, coined a phrase perfectly calculated to capture the excitement: RFID, he said, could lead to 'the Internet of Things'.[9] But the hype about RFID faded as attention moved to shiny consumer products: the smartphones, introduced in 2007 – and smart watches, smart thermostats, smart speakers and even smart cars. All these devices are sophisticated and packed with processing power – but also costly and need a substantial power source.[10]

When we debate the Internet of Things today, we usually refer not to RFID but to these devices. To some, the phrase signifies a world of over-engineered foolishness in which your toaster talks to your fridge for no good reason. Others point to security vulnerabilities: internet-enabled light bulbs that leak your passwords,[11] GPS bracelets for kids that let both parents and predators track their location[12] – and even remotely operated sex toys that reveal information about your habits that most of us might regard as rather intimate.[13]

Perhaps we shouldn't be surprised: in the age of what sociologist Shoshana Zuboff calls 'surveillance capitalism', privacy violation is now a popular business model.[14]

But amid all this hype and worry, the humble RFID continues to quietly go about its work. And my bet would be that its glory days are ahead.

Kevin Ashton's point about the Internet of Things was simple: computers depend upon data if they are to make sense of the physical world rather than just cyberspace – to track, to organise, and to optimise. Human beings have better things to do than to type in all that data – and so objects need to be built that will automatically supply that information to the computer, making the physical world intelligible in digital terms.

Many humans now carry smartphones – but physical objects do not. RFID remains an inexpensive way to keep track of them. Even if all many RFID tags do is to nod to a passing RFID reader and say 'right here, right now, this is me', that is enough for computers to make sense of the physical world: unlocking doors; keeping track of tools, components, and even medicines; automating production processes; and making small payments quickly.

RFID may not have the power and flexibility of a smart watch or a self-driving car, but it is cheap and small: cheap enough and small enough to be used to tag hundreds of billions of items. And batteries are not required – anyone who thinks that doesn't matter should remember the name of Leon Theremin.

24

The Interface Message Processor

B ob Taylor worked at the heart of the Pentagon: third floor, near the Secretary of Defense, and near, too, to the boss of ARPA, the Advanced Research Projects Agency. ARPA had been founded early in 1958, but then NASA had largely supplanted it. ARPA, in the words of *Aviation Week* magazine, was 'a dead cat hanging in the fruit closet'.[1]

Nevertheless ARPA muddled on, and in 1966, Bob Taylor and ARPA were about to plant the seed of something big.

Next to Taylor's office was the terminal room, a pokey little space where three remote-access terminals with their three different keyboards sat side by side. Each allowed Taylor to issue commands to a far-away mainframe computer: one at the University of California in Berkeley, on the other side of the continent; one at MIT in Cambridge, Massachusetts, almost 450 miles up the coast; and a Strategic Air Command mainframe in Santa Monica, called the AN/FSQ 32XD1A, or Q-32 for short.

Each of these massive computers required a different login procedure and programming language. It was, as the historians Katie Hafner and Matthew Lyon put it, like 'having a den cluttered with several television sets, each dedicated to a different channel'.[2]

Although Taylor could access these computers remotely through his terminals, they could not easily connect to each other. Nor could other ARPA-funded computers across the United States. Sharing data, dividing up a complex calculation or even sending a message between these computers was all but impossible. The next step was obvious, said Taylor. 'We ought to find a way to connect all these different machines.'[3]

Taylor talked to ARPA's boss, Charles Herzfeld, about his goal. 'We already know how to do it,' he declared, although it was not so clear that anyone really did know how to connect together a nationwide network of mainframe computers.

'Great idea,' said Herzfeld. 'Get it going. You've got a million dollars more in your budget right now. Go.' The meeting had taken twenty minutes, and Bob Taylor had better figure out how to fix the problem.[4]

Laurence Roberts of MIT had already managed to get one of his mainframes sharing data with the Q-32 over at air command in Santa Monica: two supercomputers chatting on the phone. It had been slow, fragile, and fussy.[5] Bob Taylor, Laurence Roberts and their fellow networking visionaries had something much more ambitious in mind – a network to which any computer could connect. As Roberts put it at the time, 'almost every conceivable item of computer hardware and software will be in the network.'[6]

That was an enormous opportunity; it was also a formidable challenge. Computers were rare, expensive, and puny by modern standards. The computers were typically programmed by hand by the researchers who used them. Who would persuade these privileged few to set aside their projects to write code in the service of someone else's data-sharing project? It was like asking a Ferrari owner to idle the engine in order to heat up a fillet steak, before feeding it to someone else's dog.

The solution was proposed by another computing pioneer, the physicist Wesley Clark. Clark had been following the emergence of a new breed of computer – the minicomputer, modest and inexpensive compared with the room-sized mainframes installed in universities across the United States. He suggested installing a minicomputer at every site on this new network. The local mainframe – say that hulking Q-32 in Santa Monica – would talk to the minicomputer sitting close beside it.

The minicomputer would then take responsibility for talking to all the other minicomputers on the network, and for the new and interesting problem of moving packets of data reliably around the network until they reached their destination. All the minicomputers would run in the same way, and if you wrote a networking program for one of them it would work on them all.

Adam Smith, the father of economics, would have been proud of the way that Clark was taking advantage of specialisation and the division of labour – perhaps Adam Smith's biggest idea. The existing mainframes would keep on doing what they already did well. The new minicomputers would be optimised to reliably handle the networking without breaking down. And it surely didn't hurt that ARPA could simply pay for them all.[7]

In one episode of the office comedy *The IT Crowd*, the geek heroes convince their technologically clueless boss Jen that they have 'the Internet' – it's a small box with a winking light. They offer to lend it to her as long as she promises not to break it.[8]

The beauty of Wesley Clark's idea was that, as far as any particular computer was concerned, this was pretty much how the network would appear. Each local mainframe had to be programmed merely to talk to the little black box beside it – the local minicomputer. If you could do that, you could talk to the entire network that stood behind it.[9]

The little black boxes were actually large and battleship grey. They were called interface message processors, or IMPs. The IMPs were customised from Honeywell minicomputers that were the size of refrigerators and weighed more than 400kg apiece.[10] They cost $80,000 each, more than half a million dollars in today's money.[11]

The network designers wanted message processors that would sit quietly, with minimal supervision, and just keep on working, come heat or cold, vibration or power surge, mildew, mice, or – most dangerous of all – curious graduate students with screwdrivers. Military-grade Honeywell computers seemed like the ideal starting point, although their armour plating may have been overkill.[12]

The prototype, IMP 0, emerged early in 1969. It did not work. A young engineer worked on fixing it for months, manually unwrapping and rewrapping wires on pins one-twentieth of an inch apart. It wasn't until October that year that IMP 1 and IMP 2 were in position at UCLA in Los Angeles, and the Stanford Research Institute 350 miles up the coast.

On 29 October 1969, two mainframe computers exchanged their first word through their companion IMPs. It was, biblically enough, 'LO' – the operator had been trying to type 'Login', and the network had collapsed after two letters.[13] A stuttering start, but the ARPANET had been switched on.

Other networks followed, and then a decade-long project to interconnect these into a network of networks – or simply, 'the Internet'. Eventually the IMPs were supplanted by more modern devices called routers. By the late 1980s, they were museum pieces.[14]

But the world that Laurence Roberts had predicted, in which 'almost every conceivable item of computer hardware and software will be in the network', was becoming reality. And the IMPs had shown the way.

25

GPS

What would happen if GPS stopped working?

For a start, we'd all have to engage our brains and pay attention to the world around us when getting from A to B. Perhaps this would be no bad thing: people would be less likely to drive into rivers or over cliffs through misplaced trust in their navigation devices. Pick your own favourite story about the kind of idiocy only GPS can enable: mine is the Swedish couple who misspelled the Italian island of Capri and turned up hundreds of miles away in Carpi, asking where the sea was.[1]

But these are the exceptions. Devices that use GPS usually stop us getting lost. If it failed, the roads would be clogged with drivers slowing to peer at signs or stopping to consult maps. If your commute involves a train, there'll be no information boards to tell you when to expect the next arrival. In the UK, at least, you'll have to wait for someone to open the doors – with no GPS to tell the train it's at the platform, that won't happen automatically. Phone for a taxi, and you'll find a harassed operator trying to keep track of her fleet by calling the drivers. Open the Uber app, and – well, you get the picture. Don't even think about passing the time with Pokémon Go.

With no GPS, emergency services start struggling: operators can't locate callers from their phone signal, or identify the nearest ambulance or police car. Delivery drivers take longer to bring our shopping. There are snarl-ups at ports: container cranes need GPS to unload ships. Gaps appear on supermarket shelves as 'just-in-time' logistics systems judder to a halt. Factories stand idle when their inputs don't arrive just in time, either. Farming, construction, fishing, surveying – these are other industries mentioned by a UK government report that pegs the cost of a five-day GPS outage at £5 billion.[2]

If it lasts for much longer, we might start worrying about the resilience of a whole load of other systems that might not have occurred to you if you're thinking of GPS as a location service. It is that, but it's also a *time* service. GPS, the Global Positioning System, consists of at minimum 24 satellites that all carry clocks synchronised to an extreme degree of precision. When your smartphone uses GPS to locate you on a map, it's picking up signals from some of those satellites and making calculations based on the time the signal was sent and where the satellite was. If the clocks on those satellites stray by a thousandth of a second, you'll mislay yourself by a couple of hundred miles.

So if you want incredibly accurate information about the time, GPS is the place to get it. Consider phone networks: your calls share space with others through a technique called multiplexing – data gets time-stamped, scrambled up, and unscrambled at the other end. A glitch of just a hundred-thousandth of a second can cause problems.[3] Bank payments, stock markets, power grids, digital television, cloud computing – all depend on different locations agreeing on the time. If GPS were to fail, how well, and how widely, and for how long would backup systems keep these various shows on the road? The not very reassuring answer is that nobody really seems to know.[4]

No wonder GPS is sometimes called the 'invisible utility'.[5] Trying to put a dollar value on it has become well-nigh impossible: as the author Greg Milner puts it, you may as well ask, 'How much is oxygen worth to the human respiratory system?'[6] It's a remarkable story for an invention that first won support in the US military because it could help with bombing people – and they were far from sure they needed it: 'I know where I am, why do I need a damn satellite to tell me where I am?' recalls one early champion of the typical response from colleagues.[7]

The first GPS satellite launched in 1978, but it wasn't until the first Gulf War in 1990 that the sceptics came around. As Operation Desert Storm ran into a literal desert storm, with swirling sand reducing visibility to 5 metres, GPS let soldiers mark the location of mines, find their way back to water sources, and avoid getting in each other's way. It was so obviously life-saving, and the military had so few receivers to go around, soldiers asked their families in America to spend their own money shipping over thousand-dollar commercial devices.[8]

Given the military advantage GPS conferred, you may be wondering why the US armed forces were happy for everyone to use it. The answer is that they weren't, but they couldn't do much about it. They tried, in effect, having the satellites send two signals – an accurate one for their own use, and a degraded, fuzzier one for civilians – but companies found clever ways to tease more focus from the fuzzy signals. And the economic boost was becoming ever plainer. In 2000, President Clinton bowed to the inevitable and made the high-grade signal available to all.[9]

The American taxpayer puts up the billion-odd dollars

a year it takes to keep GPS going, and that's very kind of them. But is it wise for the rest of the world to rely on their continued largesse? In fact, GPS isn't the only global navigational satellite system. There's a Russian one, too, called GLONASS – though it isn't as good. China and the European Union have well-advanced projects, called Beidou and Galileo. Japan and India are working on systems, too.[10]

These alternative satellites might help us ride out problems specific to GPS – but they might also make tempting military targets in any future conflict, and you can imagine a space war knocking everyone's systems offline. A big enough solar storm could also do the job.[11] There are land-based alternatives to satellite navigation: the main one is called eLoran, but it doesn't cover the whole world, and some countries are putting more effort than others into their national systems.[12]

One big appeal of eLoran is that its signals are stronger. By the time GPS signals have made their twenty-thousand-kilometre journey to Earth, they're extremely weak – which makes them easy to jam, or to spoof, if you know what you're doing.[13] People who are paid to think about these things worry less about the apocalyptic scenarios – waking up one day to find the whole system offline – and more about the potential for terrorists or nation states to wreak havoc by feeding inaccurate signals to GPS receivers in a certain area.[14] Engineering professor Todd Humphreys has shown that spoofing can down drones and divert superyachts.[15] He worries that attackers could feasibly fry electricity grids, cripple mobile networks or crash stock markets.[16]

The truth is that it's hard to be sure how much damage spoofing GPS signals might do. But just ask those Swedish tourists in Carpi: knowing that you're lost is one thing; being wrongly convinced that you know where you are is another problem altogether.

V

SECRETS AND LIES

26

The Movable-Type Printing Press

On Christmas Day 1438, one Andreas Dritzehn, a prosperous citizen of the city of Strasbourg, died of the plague. It was not an unusual fate at the time, but Dritzehn's death triggered a court case that continues to intrigue to this day; Dritzehn had been in a partnership to make ... well, exactly what isn't clear. Little convex metal mirrors, for sure. They were popular with pilgrims because they absorbed the divine radiance of holy relics. But the partnership was making something else, too, perhaps. Something much bigger. And despite his substantial income, the costs of the mysterious project meant that Dritzehn was up to his ears in debt.[1]

After Dritzehn's death, his irascible brothers sued his partners. The court documents that survive tell of 'a secret art', and of removing 'the pieces from the press ... so that no one can know what it is'. Dritzehn's partners were clearly concerned that this 'secret art' might be copied. The court case was settled, the surviving Dritzehn brothers received a payout, and the senior partner continued to spend money in pursuit of his 'adventure and art'. His name? Johannes Gensfleisch zur Laden zum Gutenberg.[2]

Gutenberg, of course, was working on the printing press – or, more precisely, on an entire system that would allow durable metal type to be mass-produced, flexibly rearranged, and used to print out hundreds of copies of a book in a matter of days.

It was the *system* that mattered. The idea of making letter forms and using them to stamp out characters dates back at least to the Phaistos Disc, a clay tablet found on Crete that is nearly four thousand years old. And in the year 770 the Japanese empress Shōtoku commissioned the printing of a million prayers. Since the text was brief, a single brass plate could print the entire document.[3]

But armed with the Chinese invention of paper and the European system of alphabetical writing, Johannes Gutenberg had a much more flexible printing press in mind.

The idea was in the air – a fellow called Waldvogel seems to have been working on something similar. No wonder Gutenberg was so keen to keep his idea a secret.

At the centre of Gutenberg's system was a method of mass-producing the metal type. This was essential. A single page of text would require about 3000 letter forms – it would be impossibly time-consuming to carve them all by hand.

Gutenberg was a goldsmith, well versed in the precise craft of carving punches for coins. So he and his associates intricately carved a punch for each letter out of hard metal, with the letter form sticking up in relief – easier than carving a grooved letter form. The punch would then stamp out a 'matrix' with the letter depressed into it. Finally, the matrix would be clamped into a hand-held mould, molten alloy poured in, and the metal type itself would emerge, cooling rapidly and ready to use. If the type wore out, Gutenberg could easily make more, as long as he retained either the matrices or the punches. Need a different letter? Simply clamp a different matrix into the hand mould.

Once the type was firmly fixed in a frame, Gutenberg could brush on the oil-based ink that he had developed, firmly press slightly damp paper onto the metal, and admire the results.

And what results! Gutenberg tested his machine by printing a 28-page schoolbook, but quickly moved on to a prestige project: a magnificent edition of the Bible in Latin.[4] Enea Silvio Piccolomini, the future Pope Pius II, saw some of Gutenberg's Bible in 1455. Piccolomini praised him as 'a marvellous man' and noted that 'the type was so clear it could be read without glasses, and that all copies had been sold'.[5]

But while we continue to admire the beauty of those bibles today, what was revolutionary was not the beauty or clarity, but the economics. Since Gutenberg made it possible to mass-produce writing, the price of books collapsed. The extent of this change would be hard to exaggerate. For several centuries before Gutenberg, the price of a manuscript – a hand-written book – hovered around six months' wages. Before long, it was closer to six days' wages, and by the early 1600s, six hours' wages.[6]

The output of printed material began to soar. More books were printed in the first century after the printing press than had been hand-copied in the entire pre-Gutenberg history of Europe. That was just the beginning. In the early 1400s Cambridge University's Library contained 122 books, each one a treasure. Today there are 8 million.[7]

Printing expanded the realm of ideas, raising the prestige and fame of what we'd now call thought-leaders. For example, after printing presses arrived in Italian cities around 1470, the salary of the top professors jumped from around four times the typical skilled wage to seven or eight times.[8]

It was arguably the first mass-production process – easily

pre-dating the use of machine tools to make muskets, bicycle parts and pulley blocks for sailing ships.[9]

The printing business was a new *kind* of business. For centuries, skilled trades such as weaving had been organised by guilds, which controlled who could perform the trade, and how they could perform it. But printers operated outside the guild system as for-profit firms.[10] Merchant bankers would supply the considerable up-front investment needed to make a printing press and to typeset a book – it was hard to be a printer without going into debt. Those merchants would also organise the distribution of the product, since there were no bookshops.[11]

It was a tough business. To print an illustrated bible – the product beloved of the early printers – was a vast undertaking. Many printers didn't survive the cut-throat competition. Venice, the centre of the early printing business, had twelve printing companies in 1469. Nine of them were gone in just three years.[12]

Eventually, printers figured out that it was more profitable to produce a shorter, simpler product with a lower price tag and a longer print run. Grammar books were popular – the very thing that Gutenberg had first printed to test out his system. So were pre-packaged papal indulgences. Both were reliable revenue sources. Then there were short religious polemics, such as Martin Luther's *Ninety-Five Theses*, which – so the story goes – he nailed to the door of All Saints' Church in Wittenberg, Germany in 1517.

As the historian Elizabeth Eisenstein points out, there was nothing particularly unusual about a professor of theology like Martin Luther engaging in religious argument with the Catholic Church. And church doors were a traditional place for publicity. No, what was unusual was the speed with which the printing press disseminated the rebellious ideas of Luther and his followers. Wittenberg became a one-industry town, filled with printers.[13]

Martin Luther produced a German translation of the New Testament, which was widely printed. He described printing as 'God's highest and most extreme act of grace, whereby the business of the Gospel is driven forward'.

But the pamphlets that were circulated were often anything but graceful. They were packed with vicious caricatures – for example, portraying the Pope with a wolf's head. Catholic loyalists responded with their own counter-propaganda. The religious flame war filled the pockets of the printers, sparked the Reformation and the birth of Protestant churches – and ultimately led to the catastrophe of the Thirty Years War.[14]

A revolutionary new technology rewards inflammatory rhetoric? Who would have thought it? Modern internet trolls argue that conflict brings attention, and attention brings influence – but any German living through the seventeenth century could have attested that this was not a new idea.[15]

And what of the man who started it all? According to the British Library, Johannes Gutenberg was 'the man of the millennium', and there are few others whom one could nominate for such an honour with a straight face. But even the man of the millennium struggled to make money from the printing press.[16]

Like many of the printers who followed in his footsteps, he was eager to print those glorious, ruinously expensive bibles. And Gutenberg, remember, had been accumulating debt since his partnership with Andreas Dritzehn, 17 years previously. In 1455 – the same year that the future Pope had raved about his work – he fought yet another court case with yet another business partner. This time he lost ownership of his own printing press. If only he had stuck to printing grammar books.[17]

27

Menstrual Pads

'I wish someone would tell me what Kotex is,' said one bemused young American man at a dinner party in the 1920s. Nobody would, of course. Kotex was a code word – an arcane reference to a secret man was not meant to know. Kotex was – and remains – one of the US's most popular brands of menstrual pad. But to tell you the truth, I had never heard of it.[1]

To Sharra Vostral, author of *Under Wraps*, this is unsurprising. One of the defining missions of menstrual products such as pads, tampons and cups, says Vostral, is discretion: the rest of the world simply isn't supposed to know whether a woman is menstruating or not. Not for nothing was one early brand of tampons called 'Fibs'.[2]

Not everyone approved of the wink implicit in that name. '"Fib" is a polite word for "lie",' complained one woman to market researchers. '"Fibs" suggests something nasty, secretive, unclean. If I wanted to buy tampons at a store, I would not buy "Fibs" just because of the awful name.'[3]

Nevertheless, women have had good reason to keep quiet about menstruation. It is the centre of many taboos around the world. Passages in the Old Testament of the Bible refer to menstruation as polluting, and menstrual rags as loathsome.[4]

Such ideas have been slow to fade. In 1868 the vice-president of the American Medical Association noted that female physicians could not be trusted during their monthly 'infirmity'. Five years later the American doctor and sex educator Edward Clark argued that girls should be removed from the classroom during their periods. It was too demanding to expect them to think and menstruate at the same time. The writer Eliza Duffey sharply responded that Dr Clark had no objection to women performing strenuous housework during their periods. Perhaps he just wanted to deny education to girls? Perhaps indeed.[5]

It was hardly surprising that women preferred to keep the details of their monthly cycle to themselves, using home-made approaches. Tampons have existed for thousands of years: made from wool in Rome, vegetable fibres in Indonesia, paper in Japan, grass in Africa, papyrus reeds in Egypt and ferns in Hawaii.[6] Women would fashion pads from scraps of fabric, often washing and reusing them. We now know reused pads bring the risk of infection and even cervical cancer.[7]

But in the late nineteenth century, as home-made products were replaced by manufactured commodities in other parts of life, why not in this case?

The challenge was: how to advertise and sell a product that society found unmentionable. The first recorded attempts to sell disposable pads date to the 1890s. Johnson & Johnson made and marketed 'Lister's Towels' in the US in 1896; 'Hygienic towelettes' from the German manufacturer Hartmann were advertised in Harrods in London in 1895.[8] But these products did not make much impact. It seems that most women found it cheaper, or more comfortable, or less embarrassing, to make their own sanitary towels from whatever material they had to hand.[9]

But the key technological breakthrough came during the First World War. Kimberly-Clark, a paper company, used a new material called 'cellucotton' to make bandages. Cellucotton was made of wood pulp. It was much cheaper than cotton, and far more absorbent. At the end of the war, as Kimberly-Clark was looking for new markets, they received letters from nurses explaining that they were using the cellucotton for something other than bandages.[10]

Clearly, there was a business opportunity. But it seemed risky: wasn't the product too taboo to advertise – or even to purchase? Kimberly-Clark launched anyway, settling on the mysterious name 'Kotex'. It stands for 'cotton texture', but more importantly, young men at dinner parties had no idea what 'Kotex' meant.[11]

The new product caught on fast. For decades, women had been finding some independence by taking jobs in factories and offices. Whatever Dr Edward Clark might believe, they could think and menstruate at the same time, and they needed a convenient, disposable product. To everyone's surprise, Kimberly-Clark had a hit.

The first detailed study of the growing menstrual technology market was conducted in 1927 by Lillian Gilbreth, a pioneer in applying scientific ideas from psychology and engineering to commercial problems of marketing, ergonomics and design. She noted that modern women needed to be out and about. She emphasised that women wanted a product that was discreetly packaged – it should not crackle or rustle when being unwrapped – and 'be completely invisible no matter how tight or thin their clothes might be'.[12] The product she helped design for Johnson & Johnson could even be ordered silently by handing a shop assistant a coupon that read 'one box of Modess, please'.

But while the products themselves were made to be used in secrecy, soon there was nothing secretive about the way they were marketed. The booming demand encouraged manufacturers to bombard consumers with advertisements, albeit euphemistic ones. Men may have been mystified in the 1920s; by the 1930s some felt under siege.

The future Nobel literature laureate William Faulkner complained, 'I seem to be so out of touch with the Kotex Age here that I can't seem to think of anything myself.' It takes quite a broflake to blame Kotex adverts for your writer's block, but it says something about how quickly the previously unutterable technology had entered the cultural mainstream.

The cellucotton pad was followed in the 1930s by the commercial tampon, patented in 1933 and marketed as 'Tampax'.[13] The first menstrual cup appeared soon afterwards in 1937, patented by a woman, Leona Watson Chalmers.[14]

Then came the war. Menstrual products were marketed as a way to help women participate in the war effort. One Kotex ad showed a teenager moping, her broom and mop abandoned.

'Who would have thought you'd turn out to be a deserter from a dustmop and a few dishes ... when Mom's counting on you? ... It's girls like you taking on "homework" who release a whole army of mothers for rolling bandages and selling war bonds and driving drill presses.'[15]

Come the 1950s, of course, the adverts returned to the idea of ladies of leisure in 'soft silk twill' dresses hanging around in art galleries.

These days, women spend about $3bn a year on sanitary products in the US alone.[16] They have long since become part of the cultural conversation. From a Western perspective the old sense of embarrassment is laughable – twenty-first-century adverts have mocked the tropes of an earlier age, of blue liquids in sterile laboratories interspersed with shots of women in tight white shorts riding white horses.[17]

But in many parts of the world, it's no joke. Consider the case of Arunachalam Muruganantham, a school dropout from southern India, who in 1998 decided his wife deserved hygienic, affordable pads rather than the dirty cloth she was having to use. 'I would not even use it to clean my scooter,' he said.[18]

He began experiments to produce a simple pad-making machine – something that could bring both jobs and cheap pads to women across India. His wife walked out on him. So did his widowed mother. What he was doing was simply too humiliating.

Muruganantham is now celebrated for his invention – and, yes, his wife Shanthi did come back. But his setbacks give a sense of just how powerful the stigma remains in many parts of the world.

That stigma is one reason why – according to UNESCO – one in ten girls in sub-Saharan Africa miss school during their periods.[19] Dr Edward Clark might have approved, but this is a serious matter: after falling behind, some girls drop out entirely.[20]

Stigma alone is not to blame – there's also a lack of access to clean water and lockable washrooms. And of course, there's the problem that Arunachalam Muruganantham was trying to solve: young women can't afford the menstrual products others take for granted. William Faulkner may have felt alienated by the Kotex age – but nearly a century later, many women are still waiting for that age to arrive.

28

CCTV

Peenemünde is a sandspit in northern Germany where the river Peene meets the Baltic Sea. There, in October 1942, German engineers sat in a control room watching a television screen. It showed live, close-up images of a prototype weapon on its launch pad, some 2.5 kilometres away. They counted down. On another screen, with a wide-angle view, they saw the weapon surge skywards.[1] The test had succeeded. They were looking at something that would shape the future – but perhaps not in the way they imagined.

The V-2 – the *Vergeltungswaffen*, 'vengeance weapon' – was supposed to win Hitler the war. It was the world's first rocket-powered bomb. It travelled faster than the speed of sound, so you didn't know it was coming until it exploded. But, crucially, it couldn't be targeted precisely: the V-2s killed thousands, but not enough to tip the scales of conflict.[2]

Wernher von Braun, the brilliant young engineer behind the V-2, surrendered to the Americans and helped them win the space race. If you'd told him that his rocket test would be the first step towards putting a man on the Moon, he wouldn't have been surprised. That's exactly what motivated him.[3]

But von Braun might not have anticipated that he was also witnessing the birth of another hugely influential technology. That technology was closed-circuit television, or CCTV.

The pictures in the control room were the first example of a video feed being used not for broadcasting, but for real-time monitoring, in private – over a closed circuit. The top brass at Peenemünde worked slave labourers to their deaths, but they didn't intend to join the fatalities; they invited the television engineer Walter Bruch to devise a way for them to monitor the launches from a safe distance. And that was wise, because the first V-2 they tested did indeed blow up, destroying one of Bruch's cameras.[4]

Exactly how popular Bruch's brainchild has now become is tricky to pin down. One estimate, a few years old, puts the number of surveillance cameras around the world at 245 million – that's about one for every thirty people.[5] Another reckons there'll soon be over twice that number in China alone.[6] It's clear that the market is expanding quickly, and its global leader is a company called Hikvision, which is part-owned by the Chinese government.[7]

What is China doing with all these CCTV cameras? Here's one example. Picture the scene: you're trying to cross a busy road in the city of Xiangyang. You should wait for the lights to change, but you're in a hurry, so you make a dash for it, weaving through the traffic. A few days later, you might see your photo, name and government ID number on a huge electronic billboard above the intersection, outing you as a jaywalker.[8]

But it's not just about the public shaming: surveillance cameras will feed into the country's planned 'social credit' scheme.[9] Exactly how the national system will work remains unclear, but various trials are using both public and private-sector data to score people on whether they're a good citizen.[10] You might lose points for driving inconsiderately, paying your bills late, or spreading false information.[11] Score high, and perks might include free use of public bikes; score low, and you might be banned from taking trains.[12] The aim is to incentivise desired behaviour – or, as an official document poetically puts it, to 'allow the trustworthy to roam everywhere under heaven while making it hard for the discredited to take a single step'.[13]

Perhaps this is reminding you of a certain novel published seven years after Walter Bruch pioneered the surveillance camera. In *Nineteen Eighty-Four*, George Orwell imagined life when everything is monitored – not only in public spaces, but in people's homes. Everyone who's anyone must have a 'telescreen', through which Big Brother can watch them. But there's a hint in the story that these devices were originally something people chose to buy: when the duplicitous Mr Charrington needs to give Winston a believable reason for the apparent lack of a telescreen in his spare room, he says they were 'too expensive', and 'I never seemed to feel the need of it.'[14]

That sounds like the kind of conversation I've had recently about the voice-controlled smart speakers that some of the world's largest corporations would like to sell me, so I can ask about the weather, or say 'Alexa, turn up my central heating', or automatically monitor what's in my fridge. The comic artist Zack Weinersmith sums up the value proposition:[15]

'Can I put a device in your house that perpetually listens to everything you say and do, stores that information, profits from it, and doesn't give you access to it.'

'You'd have to pay me a lot.'

'No. You'll pay us.'

'Uh . . . pass?'

'The device can figure out when you're low on Cheez Balls and drone-deliver them in 30 minutes.'

'Give me the machine!'

Devices like the Amazon Echo and Google Home have taken off because of advances in artificial intelligence – and that's the same reason behind the burgeoning demand for CCTV cameras. They always used to need human eyes and brains, and there are only so many screens one person can look at. But algorithms can read car licence plates, and they're getting better at recognising faces. And if software can watch and listen and decipher meaning, how much surveillance you can do is limited only by computing power.

Is it reasonable to feel a little queasy about this, or should we sit back and enjoy our drone-delivered Cheez Balls?

That depends in part on the extent to which we trust the entities that are surveilling us. Amazon and Google hasten to reassure us that they aren't snooping on all our conversations: the devices themselves are just smart enough to listen for when you're saying the 'wake' word – 'Alexa', or 'OK Google' – and only then do they send audio to the cloud, for more powerful servers to decipher what we want.[16]

Then we have to trust that these devices are hard to hack – for criminals and shadowy government agencies alike. That said, not everyone baulks at the thought of the state knowing more and more about our daily lives. One Chinese woman told Australia's ABC, 'If, as our government says, every corner of public space is installed with cameras, I'll feel safe.'[17]

But then there's the awkward question of how well all this tech actually works. It might appear that the Xiangyang intersection has automated face recognition, but no: the algorithms aren't yet that reliable. Government workers are sifting through the footage.[18]

It still deters – fewer people are jaywalking. That's the idea of the Panopticon again: if you think you might be being watched, you'll act as though you are. George Orwell understood that perfectly: if anyone might be an informant, you'll watch what you say – and if you'd be scared to express a thought, maybe it's best not to think it.

So CCTV might still be a long way from living up to its potential. But for those who want it to change what we do or even how we think, that might not be such an obstacle.

29

Pornography

KATE MONSTER: The internet is really really great.
TREKKIE MONSTER: For porn!
KATE MONSTER: I got a fast connection so I don't
    have to wait.
TREKKIE MONSTER: For porn!

The opening lines of 'The Internet is for Porn', a song from the Broadway musical *Avenue Q*.[1] Innocent kindergarten teacher Kate Monster is trying to celebrate the usefulness of the internet for shopping and sending birthday greetings, while her surly neighbour Trekkie Monster insists that people really value it more for more private activities.

Is Trekkie Monster right? Well, sort of – but no, not really. Credible-seeming statistics suggest that about one in seven web searches is for porn.[2] Which is not trivial – but of course it means that six in seven web searches are *not* for porn. The most-visited porn website – Pornhub – is roughly as popular as the likes of Netflix and LinkedIn. That's pretty popular, but still only enough to rank twenty-eighth in the world.[3]

But *Avenue Q* was first performed in 2003, an age ago in

internet terms, and Trekkie Monster might have been more correct back then.

When they're new, technologies often tend to be expensive and unreliable. They need to find a niche market of early adopters, whose custom helps the technology to develop. Once it is cheaper and more reliable, it finds a bigger market, and a much broader range of uses. There is a theory that pornography played this role in the development of the internet, and a whole range of other technologies. Does the theory stack up?

Since the very dawn of art, sex has been a subject. Prehistoric cave-daubers shared a muse with schoolboy doodlers, judging by their fondness for buttocks, breasts, vulvas and comically large penises.[4] Carvings of copulating couples date back at least eleven thousand years, to goat herders in Judea.[5] About four thousand years ago, a Mesopotamian artist lovingly crafted a terracotta plaque of a woman being penetrated as she sips beer through a straw.[6] A couple of millennia later, the Moche in northern Peru enjoyed depicting anal intercourse through the medium of ceramics.[7] India's Kama Sutra dates from around the same time.[8] I could go on.

But just because people used the arts and crafts to depict erotica does not mean that erotica was the driving force behind these techniques. There's no reason to think it was.

Perhaps the first communications technology we know enough about to test the theory against is Gutenberg's printing press. The theory doesn't hold water: titillating books were certainly printed, but, as we've seen, the main market for reading material was religious.[9]

A more plausible candidate, leaping ahead to the nineteenth century, is photography. Pioneering studios in Paris did a roaring trade in, uh, 'art studies', a euphemism the authorities didn't always accept. Customers were willing to pay enough to fund the technology: for a time, it cost more to buy an erotic photograph than to hire a prostitute.[10]

By the time of the next big technological breakthrough in artistic expression – the moving picture – the word 'pornography' had taken on its modern meaning. It is derived from the Greek for 'writing' and 'prostitutes', and now it means – well, 'I know it when I see it', as the American judge Potter Stewart famously said.[11] But porn didn't really drive the movie industry, for obvious reasons. Movies were expensive: You needed a big audience to recoup your costs. That meant public viewings. And while many people paid to look at dirty pictures in the privacy of their home, far fewer people were comfortable watching a dirty movie in a public theatre.[12]

One solution came in the 1960s with the peep-show booth – an enclosed space where you'd put coins in a slot to keep a movie playing. One booth could bring in several thousand dollars a week.[13]

But the real privacy breakthrough came thanks to the video-cassette recorder, or VCR. In his book *The Erotic Engine*, writer Patchen Barss argues that it was with the VCR that porn 'came into its own as an economic and technological powerhouse'.[14]

At first, VCRs were a hard sell: they were pricey, and they came in two incompatible formats – VHS and Betamax. Who would risk plunging a significant chunk of cash into a device that might soon be obsolete? People who really wanted to watch adult movies at home, that's who. In the late 1970s, more than half of videotape sales were pornographic. Within a few years, the technology was more affordable for people who wanted to watch family movies – the market expanded, and porn's share of it shrank.[15]

A similar story can be told about cable television – and, yes, the internet. Older listeners might remember when getting online meant coaxing a dial-up modem into establishing a connection, then fretting about long-distance phone charges as it slowly chugged through a file that would nowadays download in the snap of a finger. What would motivate an ordinary person to persevere? You've guessed it. One 1990s study of Usenet discussion groups claimed that five in six images shared were pornographic.[16] A few years later, research into internet chatrooms found a similar proportion of activity devoted to sex.[17]

So in those days, Trekkie Monster wasn't far wrong. And, as he suggests to Kate, appetite for porn helped to drive demand for faster connections – better modems and higher bandwidth. It spurred innovation in other areas, too: online porn providers were pioneers in web technologies such as video file compression and user-friendly payment systems, and also in business models, such as affiliate marketing pro-grammes.[18] All these ideas went on to find much wider uses, and as the internet expanded it gradually became less for porn, and more for all that other stuff.

Nowadays the internet is making life hard for professional pornographers. Just as it's hard to sell a newspaper subscription or a music video when so much is available free online, it's hard to sell porn when sites like Pornhub are giving it away. Much of this free porn is pirated, and it's an uphill struggle to get the illegally uploaded content removed.[19] The travails of porn producers are chronicled by Jon Ronson in his podcast series The Butterfly Effect.[20] One emerging niche is pro-ducing 'custom' porn, for clients such as the man who pays to see beautiful women contemptuously destroy his stamp collection.[21]

But, of course, what's bad for the content creators is good for the aggregator platforms, which make their money through advertising and premium subscriptions. The big player in porn at the moment is a company called Mindgeek. It owns not only Pornhub, but several other top porn websites.[22]

In *Avenue Q*, Trekkie Monster appears to do nothing all day but surf for porn, so the other characters are surprised when he reveals he's a multi-millionaire. As he explains: 'In volatile market, only stable investment is . . . porn!'[23]

And, once again, Trekkie Monster is nearly right, but not quite. For sure, there's money in porn. But the best way to make it may be to invest in the technologies that enable it and that it enables – in the past, in Parisian photo studios, or companies making VCRs or high-speed modems; today, in Mindgeek's algorithms that suggest content and keep eyeballs on screens. And what will Trekkie Monster be singing in the future? 'Robots are for porn', perhaps.[24] The role of sex in accelerating technology is unlikely to be finished yet.

30

Prohibition

Economists have an image problem. People think we shamelessly massage statistics, we overconfidently make terrible predictions, we're no fun at drinks parties – that kind of thing. Perhaps some of the blame lies with the man who, a century ago, was probably the most famous economist in the world. His name was Irving Fisher.

It was Fisher who notoriously claimed, in October 1929, that the stock market had reached 'a permanently high plateau'; nine days later came the crash that led to the Great Depression.[1] As for parties, the best that can be said for Fisher was that he was a generous host. One dinner guest wrote, 'While I ate right through my succession of delicious courses, he dined on a vegetable and a raw egg.' Fisher was a fitness fanatic who avoided meat, tea, coffee and chocolate.[2]

Fisher, of course, didn't drink alcohol. But he also thought nobody else should drink alcohol either. So, it seems, did the entire economics profession: Fisher claimed to have been unable to find a single economist willing to oppose prohibition in a debate.[3]

Prohibition was America's ill-fated attempt to outlaw the

manufacture and sale of alcohol. It began in 1920. It was a remarkable reform – the country's fifth-largest industry was suddenly illegal.[4] Fisher predicted it would 'go down in history as ushering in a new era in the world, in which accomplishment this nation will take pride forever'.[5]

That worked out about as well as the permanently high plateau: historians have typically seen prohibition as a farce.[6] It was so widely flouted that consumption of alcohol decreased by only about a fifth.[7] It finally ended in 1933 when one of Franklin D. Roosevelt's first acts as president was to re-legalise beer, bringing cheering crowds to the White House gates.[8]

The roots of prohibition are generally traced to religion, perhaps laced with class-based snobbery.[9] But economists had another concern: productivity. Wouldn't sober nations outcompete those with a workforce of drunks? Fisher worried about absenteeism and 'Blue Mondays', the hangover from a weekend binge.[10]

Fisher seems to have happily taken some liberties with figures. He claimed, for example, that prohibition was worth $6 billion to America's economy. Was this the result of careful study? No, said one bemused critic: Fisher evidently started with reports from a few individuals that a stiff drink on an empty stomach made them 2 per cent less efficient; then he assumed that workers habitually downed five stiff drinks just before work; then he multiplied the two by five, and concluded that alcohol lopped 10 per cent off production.[11] Dubious, to say the least.

Economists might have been less surprised by the failure of prohibition if they'd been able to fast-forward half a century to Gary Becker's insights on 'rational crime'.[12] Becker explained that making something illegal simply adds a cost that rational people will weigh up alongside other costs and benefits: that cost is the penalty if you're caught, modulated by the probability of being caught. Becker meant it, too: the first time I met him, he parked in such a way as to risk getting a ticket. 'I don't think they check that carefully,' he told me, cheerfully admitting that he had committed a rational crime.[13]

'Rational criminals', said Becker, will supply prohibited goods at the right price. Whether consumers will pay that price depends on what economists call elasticity of demand. Imagine, for example, that the government bans ... broccoli. Would black marketeers grow broccoli in secluded back gardens and sell it down dark alleys for an inflated price? It's unlikely, because demand for broccoli is elastic – hike the price, and most of us will buy cauliflower or cabbage instead.

With alcohol, it turns out, demand is inelastic: hike the price, and many will still pay. Prohibition was a boon for rational criminals like Al Capone, who defended his boot-legging in entrepreneurial terms: 'I give the public what the public wants. I never had to send out high pressure salesmen. Why, I could never meet the demand.'[14]

Any rational criminal would want to reduce their chance of being caught, and one approach is to bribe the authorities. An investigation in Philadelphia in 1928 turned up numerous police officers who had mysteriously accumulated savings some 50 to 80 times their annual salaries; one hopefully claimed he'd been lucky at poker.[15]

Black markets change incentives in other ways. Your competitors can't take you to court, so why not use whatever means necessary to establish a local monopoly? Mob violence may have spiked after prohibition; that belief was certainly one reason why prohibition was repealed.[16] Every shipment of illegal goods carries some risk, so why not save space by making your product more potent? During prohibition, consumption of beer declined relative to spirits; when it ended, that reversed.[17]

And why not cut costs by lowering quality? If you're making 'moonshine' – strong, illegal drink – you don't have to list your ingredients on the label. There's dispute about how prohibition affected productivity, but one employer complained that 'the stuff available to labor, and there is plenty of it, is so rotten that it takes the drinking man two or three days to get over his spree'.[18] Not so much eliminating Blue Monday, as extending it into Tuesday or Wednesday.

America wasn't the only country to try prohibition – others included Iceland, Finland and the Faroe Isles – but nowadays nations that strictly ban booze tend to be Islamic.[19] Others have partial restrictions. In the Philippines, for instance, you can't buy alcohol on election day,[20] or in Thailand on Buddhist holidays – except at the airport duty-free.[21] America still has 'dry' counties,[22] and local 'blue laws' that ban sales on Sundays.[23]

Those laws inspired the economist Bruce Yandle to coin a term that's become common in the branch of economics called public-choice theory: 'bootleggers and Baptists'.[24] The idea is that regulations are often supported by a surprising alliance of noble-minded moralists and profit-driven cynics.

Think about bans on cannabis. Who supports them? The 'Baptists' are anyone who thinks cannabis is wrong; the 'bootleggers' are the rational criminals who profit from illicit dope, along with anyone else with an economic interest in anti-drugs laws, such as the bureaucrats paid to enforce them.[25]

In recent years, that alliance has weakened: cannabis has been legalised or decriminalised from California to Canada, from Austria to Uruguay.[26] Debates in other countries are raging: if you're going to impose costs on cannabis producers, should you do that by trying to enforce laws against selling cannabis, or by making it legal and imposing a tax?

In the UK, the free-market think tank the Institute for

Economic Affairs crunched the numbers on elasticity of demand: they reckon a 30 per cent tax would almost eradicate the black market, raise about £700 million for the government, and lead to safer drugs too, just as the end of prohibition led to safer drinks.[27]

Today you'd have no trouble finding economists to oppose prohibition of cannabis: five Nobel memorial prize winners have called for an end to the 'war on drugs', and instead for 'evidence-based policies underpinned by rigorous economic analysis'.[28]

Naturally, that evidence covers productivity: some studies find that cannabis impairs functioning; others find no effect; one outlier even found that toking a spliff gave a short-term boost to workers' hourly output.[29] One wonders what Irving Fisher would have made of that.

31

'Like'

Leah Perlman draws comics, sharing her ideas on topics such as 'emotional literacy' and 'self-love'. When she started to post them on Facebook, she discovered that her friends found them 'healing and endearing'.[1]

But then Facebook changed its algorithm – how it decides what to put in front of our eyeballs. If social media is a big part of your life, an algorithm change can come as a shock: you might suddenly find that your content is being shown to fewer people.

That is what happened to Leah. Her comics started to get fewer likes. She told an interviewer for Vice.com that it felt like she wasn't getting enough oxygen. She could pour her heart and soul into a drawing, then watch as it racked up only 20 likes.[2]

It's easy to empathise. Social approval can be addictive, and what's a Facebook 'like' if not social approval distilled into its purest form? Researchers now liken our smartphones to slot machines. They trigger the same reward pathways in our brain: more likes, new notifications, even an old-fashioned email – we never know what we'll get when we pull the lever.[3]

Faced with a sudden drop in her likes, Leah started to buy ads on Facebook – that is, she started to pay Facebook so more people would see her comics. She just wanted the attention; but she felt embarrassed admitting it.[4] In 2016 she hired a social media manager to deal with Facebook for her. She just didn't want the anxiety.[5]

There's an irony behind Leah's embarrassment. Before she was a comic artist, Leah was a developer at Facebook. In July 2007, her team invented the Like button.

It's now ubiquitous across the web, as content creators invite you to signal your approval to your Facebook friends. There are similar features everywhere from YouTube to Twitter. For the platforms, the benefit is obvious – a single click is the simplest possible way to get users to engage, much easier than typing out a comment. But the idea wasn't immediately appreciated: Facebook's CEO, Mark Zuckerberg, repeatedly knocked it back. There were debates about the wording – 'like' was nearly 'awesome'.[6] And the symbol: while a thumbs-up means approval in most cultures, in some it has a rather cruder and less friendly meaning.[7]

Eventually, in February 2009, the Like button launched. Leah Perlman remembered how quickly it took off. Almost immediately, 50 comments would become 150 likes. More engagement, more status updates, more content. 'It all just worked.'[8]

Meanwhile, at Cambridge University, Michal Kosinski was doing a PhD in psychometrics – the study of measuring psychological profiles. A fellow student had written a Facebook app to test the 'big five' personality traits: openness, conscientiousness, extraversion, agreeableness and neuroticism. Taking the test gave the researchers permission to access your Facebook profile – with your age, gender, sexual orientation and so on. The test went viral. The dataset swelled to millions of people. And the researchers could see every time those people had clicked 'like'.[9]

Kosinski realised he was sitting on a treasure trove of potential insights. It turned out, for example, that a slightly higher proportion of gay men than straight men 'liked' the cosmetics brand MAC. That's only one data point; Kosinski couldn't tell if you're gay from a single like. But the more likes he saw, the more accurate guesses he could make – at your sexual orientation, religious affiliation, political leanings, and more. Kosinski concluded that if you'd liked 70 things, he would know you better than your friends; after 300 likes, he knew you better than your partner.[10]

Facebook has since restricted what data gets shared with app developers like Kosinski's colleague.[11] But one organisation still gets to see all your likes and more besides: Facebook itself.[12] And it can afford to employ the world's brightest machine-learning developers to tease out conclusions.

What can Facebook do with its window into your soul? Two things. First, it can tailor your newsfeed so you spend more time on Facebook – whether that means showing you cat videos, inspirational memes, things that will outrage you about Donald Trump, or things that will outrage you about Donald Trump's opponents. This isn't ideal: it makes it harder and harder for people with different opinions about Donald Trump to conduct a sensible conversation.

Second, it can help advertisers to target you. The better the ads perform, the more money it makes.

Targeted adverts are nothing new. Long before the internet and social media, if you were opening a new bicycle shop in Springfield, say, you might have chosen to advertise in the *Springfield Gazette* or *Cycling Weekly*, rather than the *New York Times* or *Good Housekeeping*. Of course, that still wasn't very efficient: most *Gazette* readers wouldn't be cyclists, and most subscribers to *Cycling Weekly* wouldn't live near Springfield. But it was the best you could do.

You could say that Facebook simply improves that process, and it's nothing to worry about. If you ask it to show your ads only to Springfield residents who've liked content on cycling, who could object to that? This is the kind of example Facebook tends to cite when it defends the concept of 'relevant' advertising.[13] But there are other possible uses that might make us feel more queasy. How about advertising a house for rent, and not showing that advert to African Americans? The investigative website ProPublica wondered if that would work; it did. Facebook said oops, that shouldn't have happened – it was a 'technical failure'.[14]

Or how about helping advertisers to reach self-proclaimed 'Jew haters'? ProPublica showed that was possible, too; Facebook said oops, it wouldn't happen again.[15] This kind of thing might worry us because not all advertisers are as benign as bicycle shops – you can also pay to spread political messages, which may be hard for users to contextualise or verify. A firm called Cambridge Analytica claimed it had swung the 2016 election for Donald Trump, in part by harnessing the power of the Like button to target individual voters[16] – much to the horror of Michal Kosinski, the researcher who had first suggested what might be possible.[17]

What about the idea of helping unscrupulous marketers to pitch their products to emotionally vulnerable teenagers at moments when they're feeling particularly down? In 2017, *The Australian* reported on a leaked Facebook document apparently touting just this ability.[18] Facebook said oops, there'd been an 'oversight', and it 'does not offer tools to target people based on their emotional state'.[19] Let's hope not, especially as Facebook has previously admitted to manipulating people's emotional states by choosing whether to show them sad or happy news.[20]

In reality, Facebook's potential for mind control still seems to be reassuringly limited. Experts who've looked into Cambridge Analytica question how effective they really were.[21] And for all the targeting, analysts report that the click-through rate on Facebook adverts still averages less than 1 per cent.[22]

Perhaps we should worry more about Facebook's undoubted proficiency at serving us more adverts by sucking in an inordinate amount of our attention, hooking us to our screens. How should we manage our compulsions in this brave new social media world? We might cultivate emotional literacy about how the algorithm affects us; and if social approval feels as vital as oxygen, maybe more self-love is the answer. If I see any good comics on the subject, I'll be sure to click 'like'.

V

WORKING TOGETHER

32

Cassava Processing

In 1981, in Nampula, Mozambique, a young Swedish doctor named Hans Rosling was puzzled. More and more people were coming to his clinic suffering from paralysis in their legs. Could it be an outbreak of polio? No. The symptoms were not in any textbook. His puzzlement turned to alarm. With Mozambique slipping into a civil war, might it be chemical weapons? He packed his wife and young children off to safety and continued his investigations.[1]

The resolution of the mystery sheds light not just on paralysis of the legs, but on one of the biggest economic questions – why do humans have an economy at all?

Let's return to Mozambique in due course. First, an outback adventure. In 1860, Robert Burke and William Wills led the first European expedition across the interior of Australia. Burke, Wills and their companion John King ran out of food on the return journey. They became stranded at a stream called Cooper's Creek, unable to carry enough water to cross a stretch of desert to the nearest colonial outpost at the unpromisingly named 'Mount Hopeless'.[2]

William Wills wrote, 'We have been unable to leave the

creek. Both camels are dead and our provisions are done. We are trying to live the best way we can, like the Blacks, but find it hard work.'[3]

By 'the Blacks', Wills meant the local Yandruwandha people, who seemed to thrive despite conditions that were proving too tough for Burke, Wills and King. The Yandruwandha gave the explorers cakes made from the crushed seed pods of a clover-like fern called nardoo – but later Burke fell out with them and, unwisely, drove them away by firing his pistol.[4]

But perhaps Burke, Wills and King had already learned enough to survive? They found fresh nardoo and decided to make their own cakes. At first, all seemed well. The nardoo cakes satisfied their appetites, yet they felt ever weaker. Wills wrote that the nardoo, 'Will not agree with me in any form ... The stools it causes are enormous ... '

Within a week, he and Burke were dead.[5]

It turns out that safely preparing nardoo is a complex process. Nardoo is packed with an enzyme called thiaminase. Thiaminase breaks down the body's supply of Vitamin B1, preventing the body using the nutrients in food. Burke, Wills and King were full, but starving.[6]

The Yandruwandha roasted the nardoo spores, ground the flour with water, and exposed the cakes to ash – each step making the thiaminase less toxic. It is not something one learns to do at a glance.[7] Barely alive, John King threw himself on the mercy of the Yandruwandha, and they took pity on him, keeping him alive until European help arrived months later.

As a foodstuff, nardoo is a curiosity. The same cannot be said of cassava roots, which are a vital source of calories in many tropical countries, particularly for subsistence farmers in Africa.[8] But cassava is toxic, like nardoo. And like nardoo it requires a tedious and complex preparation ritual to make it safe. The cassava root will otherwise release hydrogen cyanide, the same active ingredient as Zyklon B, the gas used in the Third Reich's death camps.[9]

What makes cassava particularly treacherous is that while some processing will reduce the bitter taste and the risk of immediate cyanide poisoning, only the full, time-consuming ritual can guarantee that you won't be slowly poisoned, producing a condition called Konzo, with symptoms including . . . sudden paralysis of the legs.[10]

An epidemiologist named Julie Cliff eventually figured out that that's what had happened to the patients at Hans Rosling's clinic in Mozambique.[11] Their meals of cassava had been incompletely processed. Already hungry and malnourished, they could not wait long enough to make the cassava safe.[12]

Toxic plants are everywhere. Often, simply cooking makes them edible. But how does anyone learn the elaborate preparation needed for cassava or nardoo?

Joseph Henrich, an evolutionary biologist, has an answer: no single person does. This knowledge is cultural. Our cultures evolve through a process of trial and error analogous to evolution in biological species.

Like biological evolution, cultural evolution can – given enough time – produce impressively sophisticated results. Somebody stumbles on one step that seems to make cassava less risky; that spreads and another step is discovered. Over time, complex rituals can evolve, each slightly more effective than the last.

In the Amazon, where humans have eaten cassava for thousands of years, tribes have learned the many steps needed to detoxify it completely: scrape, grate, wash, boil the liquid, leave the solid to stand for two days, then bake. Ask why they do this, and they won't mention hydrogen cyanide. They'll simply say 'This is our culture.'

In Africa, cassava was introduced only in the seventeenth century. It didn't come with an instruction manual.[13] Cyanide poisoning is still an occasional problem; people take shortcuts because cultural learning is still incomplete.[14]

'Cultural evolution,' writes Henrich, 'is often much smarter than we are.'[15]

Whether it's constructing an igloo, hunting an antelope, lighting a fire, making a longbow or processing cassava, we learn not by understanding from first principles but by imitating. One study challenged participants to place weights on the spokes of a wheel to maximise the speed at which it rolled down a slope. Each person's best effort would be passed to the next person. Because they benefited from earlier experiments, later participants did much better. Yet when asked, they showed no sign of actually understanding why some wheels rolled faster than others.[16]

Other studies show that the verb 'to ape', meaning to copy, is ironically misplaced: the only ape with the instinct to imitate is us. Tests reveal two-and-a-half-year-old chimps and humans have similar mental capacities – unless the challenge is to learn by copying someone. The toddlers are vastly better at copying than the chimps.[17]

And humans ritualistically copy in a way that chimps do not. When a demonstrator solves a puzzle but includes redundant movements, chimps will generally eliminate the superfluous activity, but humans – both children and adult – will slavishly copy the demonstration, including pointless steps. Psychologists call this 'over-imitation'.[18]

It may seem like the chimps are the smart ones here. But if you're processing cassava roots, over-imitation is exactly what you should be doing. If Henrich is right, human civilisation is based less on raw intelligence than on a highly developed ability to learn from each other.[19] Over the generations our ancestors accumulated useful ideas by trial and error, and the next generation simply copied them. Yes, we now have the scientific method – but we shouldn't look down on the kind of collective intelligence that saved John King's life. It's what made civilisation – and the economy – possible.

33

Pensions

'I customarily killed old women ... they all died, there by the big river ... I didn't used to wait until they were completely dead to bury them ... The women were customarily afraid of me.'

No wonder. That's a man from the Aché, an indigenous tribe in eastern Paraguay, talking to anthropologists Kim Hill and Magdalena Hurtado. He explained that grandmothers helped with chores and babysitting, but when they got too old to be useful you couldn't be sentimental. The usual method was an axe to the head. For the old men, Aché custom dictated a different fate. They were sent away, and told never to return.[1]

What obligations do we owe to our elders? It's a question as old as humankind. And the answers have varied widely, at least if surviving traditional societies are any guide. Jared Diamond, another anthropologist, says the Aché are hardly outliers. Among the Kualong, in Papua New Guinea, when a woman's husband died, it was her son's solemn duty to strangle her. In the Arctic, the Chukchi encouraged old people to kill themselves with the promise of rewards in the afterlife.[2]

Yet many tribes took a very different approach: they were

gerontocracies, in which the young do as the old say. Some even expected adults to pre-chew food for their aged and toothless parents.[3]

What does seem common is the expectation that, until your body let you down completely, you'd keep working.[4] That's no longer true. Many of us expect to reach a certain age, then receive money from the state or our former employers, not in return for work today, but in recognition of our work in the past. This curious stage of life is called 'retirement', and the payment, a 'pension'.

Pensions for soldiers date back at least as far as ancient Rome – the word 'pension' comes from the Latin for 'payment'. But only in the nineteenth century did they spread far beyond the military.[5] The first universal state pension came in Germany in 1890.[6]

The right to support in old age is still far from global – nearly a third of the world's older people have no pension,[7] and for many of the rest the pension is not enough to live on. Still, in many countries, generations have grown up assuming they'll be well looked after in old age.

It's becoming a challenge to meet that expectation. For years, economic policy wonks have been sounding the alarm about a slow-burn crisis in the pension system.[8] The problem is demographic. Half a century ago, in the OECD – a club of rich nations – the average 65-year-old woman could expect to live about 15 more years. Today, she can expect at least 20.[9] Meanwhile, families have shrunk from 2.7 children to 1.7: the pipeline of future workers is drying up.[10]

All that has many implications, some good and some bad. But for pensions, the situation is stark: there'll be many more retirees to support, and many fewer workers paying taxes to support them. In the 1960s, the world had nearly 12 workers for every older person; today, it's under eight; by 2050, it'll be just four.[11]

Both state and private pension systems now look expensive. Employers have been scrambling to make theirs less generous. Forty years ago, most American workers were on so-called 'defined-benefit' plans, which specify what you'll get when you retire. Now it's fewer than one in ten.[12]

The new norm, 'defined contribution', specifies what your employer will pay into your pension pot rather than what income you'll be able to get out of it. Such pensions don't logically have to be more miserly than defined benefit schemes – but they usually are, often vastly so.

It's easy to understand why employers are ditching defined benefits: pension promises can prove expensive to keep. Ponder the case of John Janeway, who fought in the US Civil War. His military pension included benefits for a surviving spouse when he died. When Janeway was 81, he married an 18-year-old. The army was still paying Gertrude Janeway her widow's pension in 2003, nearly 140 years after the war ended.[13]

The wonks can see trouble ahead: a bulge of workers is approaching retirement and their workplace pensions may be worth less than they'd expected. That's why governments around the world are trying to persuade individuals to save more towards their old age.[14]

But it's not easy to get people to focus on the distant future. One survey finds that under-50s are barely half as likely as over-50s to say retirement is their top financial concern.[15] When you're saving for your first house, or raising a young family, you may not feel a pressing need to provide for the old person you'll one day become. Indeed, you may find it hard to conceive of that future old person as you. It's a mental block summed up by Homer Simpson: 'That's a problem for Future Homer. Man, I don't envy that guy.'[16]

Behavioural economists have come up with some clever

solutions, such as automatically enrolling people in workplace pension schemes, and scheduling more saving from future pay rises. These 'nudges' work pretty well – we could opt out, but instead we tend to save through sheer inertia.[17]

But they don't solve the fundamental demographic problem. No amount of saving changes the fact that we'll always need current workers to generate the wealth to support current pensioners – whether that's through paying taxes, renting properties owned by retirees, or working for companies in which pension funds are the major shareholders.

Some think we'll need a more radical shift in our attitudes to old age. There's talk of retirement itself being 'retired'.[18] Perhaps, like our ancestors, we'll be expected to work for as long as we're able.

But the varied customs of ancestral societies should give us pause, because they appear to have evolved in response to some discomfitingly hard-nosed trade-offs. Whether elders could expect lovingly pre-chewed food or an axe by the big river seems to have depended on whether the benefits they offered to the tribe outweighed the costs of supporting them. In tribes like the Aché, those costs were higher – because they moved around a lot, or food was frequently scarce.[19]

Today's societies are rich and sedentary by comparison: we can afford the rising cost of pensions, if we choose. But there are other differences, too. Once we relied on elders to store knowledge and instruct the young. Now, knowledge dates quickly – and who needs Grandma when we have schools and Wikipedia?

We might hope we're long past the days when levels of respect for old people unconsciously tracked some balance of costs and benefits. Still, if we believe that a dignified old age is a right, perhaps we should be saying that, as clearly and as often as possible.

34

QWERTY

It isn't easy to type QWERTY on a QWERTY keyboard. My left-hand little finger holds the shift key, then the other fingers crab sideways across the upper row. Q-W-E-R-T-Y. That particular combination is awkward. There's a lesson here: it matters where the keys sit on your keyboard. There are good arrangements, and bad ones.

Many people think that QWERTY is a bad one – in fact, that it was deliberately designed to be slow and awkward. Could that be true? And why do economists, of all people, argue about this? It turns out that the stakes here are higher than they might first appear.

But let's start by figuring out why anyone might have been perverse enough to want to slow down typists. In the early 1980s, I persuaded my mother to take down her mechanical typewriter from a high shelf, delighted by the idea that I could transcend my awful handwriting using this miraculous machine.

I'd bang down on a key – not easy for little fingers – and when I did, a lever would flick up from behind the keyboard, a tiny golf club of a thing that would whack hard against an inked ribbon, squeezing that ink against a sheet of paper. On the end of the lever – called a type bar – would be a pair of reversed letters in relief. I discovered through impish trial and error that if I hit several keys at once, the type bars all flew up at the same time into the same spot, like two or three golfers all trying to strike the same ball. For a nine-year-old boy this was fun. For a professional typist, the results would leave something to be desired.

And a professional typist might just run into that problem. Typing at 60 words a minute – no stretch for a good typist – means five or six letters striking the same spot each second. At such a speed, the typist might need to be slowed down for the sake of the typewriter. That is what QWERTY supposedly did.

Then again, if QWERTY really was designed to be slow, how come the most popular pair of letters in English, T–H, are adjacent and right under the index fingers? The plot thickens.

The father of the QWERTY keyboard, Christopher Latham Sholes, a printer from Wisconsin, sold his first typewriter in 1868 to Edward Payson Porter of Porter's Telegraph College, Chicago – which gives a clue to what was going on. The QWERTY layout was designed for the convenience of telegraph operators transcribing Morse code – that's why, for example, the Z is next to the S and the E, because Z and SE are indistinguishable in American Morse code. The telegraph receiver would hover over those letters, waiting for context to make everything clear.[1]

So the QWERTY keyboard wasn't designed to be slow. But it wasn't designed for the convenience of you and me, either. Why do we still use it?

The simple answer is that QWERTY won a battle for dominance in the 1880s. Sholes's design was taken up by the gunsmiths E. Remington and Sons. They finalised the layout and put it on the market for $125 – perhaps $3000 in today's money, and many months' income for the secretaries who would have used the machine.[2]

It wasn't the only typewriter around – Sholes has been described as the '52nd man to invent the typewriter' – but the QWERTY keyboard emerged victorious. The Remington company had been canny about providing typing courses on QWERTY, and when it merged with four major rivals in 1893, they all adopted QWERTY as what became known as 'the Universal layout'.[3]

Yet this brief struggle for market dominance in 1880s America determines the layout of a keyboard on an iPad. Nobody *then* was thinking about our interests *today* – but their actions control ours. These things have a momentum of their own.

And that's a shame, because more logical layouts exist: notably the Dvorak, designed by August Dvorak and patented in 1932. It favours the stronger hand (left- and right-hand layouts are available) and puts the most-used keys together. The US Navy conducted a study in the 1940s demonstrating that the Dvorak was vastly superior: training typists to use the Dvorak layout would pay for itself many times over.

So why didn't we all switch to Dvorak? The problem lay in coordinating the switch. QWERTY had been the universal layout since before August Dvorak was born. Most typists trained on it. Any employer investing in a costly typewriter would naturally choose the layout that most typists could use. Economies of scale kicked in: QWERTY typewriters became cheaper to produce, and thus cheaper to buy. Everyone trained on QWERTY; every office used it. Dvorak keyboards never stood a chance.

So now we start to see why this case matters. For a leading

economic historian, Paul David, QWERTY is the quintessential example of something economists call 'lock-in'. Paul David argued that we get locked into standards like QWERTY all the time.

This isn't about typewriters. It's about Microsoft Office and Windows, Amazon's control of the online retail link between online buyers and sellers, and Facebook's dominance of social media. If all your friends are on Facebook apps such as Instagram and WhatsApp, doesn't that lock you in as surely as a QWERTY typist? It doesn't matter if you, personally, might want to make the shift: you can't do it by yourself.

The stakes here are high: lock-in is the friend of monopolists, the enemy of competition, and may require a robust response from regulators.

But there are two sides to the argument. Maybe these dominant standards are dominant not because of lock-in, but just because the alternatives simply aren't as compelling as we imagine. Consider the famous Navy study that demonstrated the superiority of the Dvorak keyboard. Two economists, Stan Liebowitz and Stephen Margolis, unearthed that study, and concluded it was badly flawed. They also raised an eyebrow at the name of the man who supervised it – the Navy's leading time-and-motion expert, one Lieutenant-Commander . . . August Dvorak.[4]

Liebowitz and Margolis don't deny that the Dvorak design may be better. After all, the world's fastest alphanumeric typists do use Dvorak keyboards. They're just not convinced that this was ever an example where an entire society was desperate to switch to a hugely superior standard yet unable to coordinate. These days, most of us peck away at our own emails, on devices that make it easy to switch your keyboard layout. Windows, iOS and Android all offer Dvorak layouts. If you prefer it, you no longer need to persuade your co-workers, other employers and secretarial schools to switch with you. You can just use it. Nobody else is even going to notice.

Yet most of us stick with QWERTY. The door is no longer locked, but we can't be bothered to escape.

Lock-in seems to be entrenching the position of some of the most powerful and valuable companies in the world today – including Apple, Facebook and Microsoft. Maybe those locks are as unbreakable as the QWERTY standard once seemed. Or maybe they are vulnerable to being crow-barred off at the first sign of restless consumers: it wasn't long ago, after all, that people worried about users being locked in to MySpace.[5] One of the most important questions in the economy today is whether the locks that surround technology standards are formidable . . . or feeble.

35

The Langstroth Hive

It's a little known fact that economists love bees – or at least, the idea of bees. The Royal Economic Society's logo is a honeybee. A famous work of proto-economics, *The Fable of the Bees*, published in 1732 by a Dutch-born Londoner named Bernard Mandeville, uses honeybees as a metaphor for the economy, and anticipates modern economic concepts such as the division of labour and the invisible hand.[1]

And when a future winner of the Nobel memorial prize in economics, James Meade, was looking for an example of a tricky idea in economic theory, it was to the honeybee that he turned for inspiration.

The tricky idea was what economists call a 'positive externality'. It's like a kind of topsy-turvy pollution: something good that a free market won't produce enough of, meaning that the government might want to subsidise it.

For James Meade, the perfect example of a positive externality was the relationship between apples and bees. Imagine, wrote Meade in 1952, a region containing some orchards and some beekeeping. If the apple-farmers planted more apple trees, the beekeepers would benefit, because that would mean more honey. But the apple farmers wouldn't enjoy that benefit, that positive externality, and so they wouldn't plant as many apple trees as would be best for everyone. This was, according to Meade, 'due simply and solely to the fact that the apple-farmer cannot charge the bee-keeper for the bees' food'.[2]

Close your eyes and you can see Meade's example coming to life: the haze of early summer, the cool shade of the apple trees, the buzz of the busy bees. No wonder the example has been handed down the years. It's vivid, evocative – and entirely mistaken. Apple blossom produces almost no honey. And that's only the first thing James Meade didn't know about bees.

To understand Meade's more fundamental error, we need a brief history of humans and honeybees. Once upon a time, there was no beekeeping – only honey hunting, trying to steal honeycombs from wild bees. We see this depicted in cave paintings.[3]

Then came beekeeping, at least five thousand years ago.[4] The Greeks, the Egyptians and the Romans were all partial to domesticated honey. By the Middle Ages, beekeepers were using skep hives: they're the classic woven beehives that look like a tapering stack of straw tyres.

The trouble with skep hives is that if you want to get the honey, you need to get rid of the bees – and beekeepers would generally poison them with sulphurous smoke, shake them off, scoop out the honey, and worry about getting another bee colony in due course. People started to worry about this waste and disdain for a creature that not only gives us honey but pollinates our plants. A bee-rights movement emerged in the US in the 1830s, with the motto 'Never kill a bee'.

It was in everyone's interests – especially that of the bees – to build a better beehive. And in 1852, the US Patent Office awarded patent number 9300A to an American clergyman named Lorenzo L. Langstroth for a movable-frame beehive – often known today simply as the Langstroth hive.[5]

The Langstroth hive is a wooden box with an opening at the top, and frames that hang down – carefully separated from each other by the magic gap of 5/16th of an inch (or 8 millimetres) – any smaller, or larger, and the bees start adding their own inconvenient structures. The queen is at the bottom, confined by a 'queen excluder' – a mesh that prevents her entry but permits worker bees through. This keeps her bee larvae out of the honeycombs. The honeycombs are easily pulled out and harvested by a spinning centrifuge that flings out, filters and collects the honey. The Langstroth hive, a marvel of design and efficiency, allowed the industrialisation of the bee.[6]

It's this industrialisation that James Meade hadn't quite grasped. The honeybee is a thoroughly domesticated animal. With Langstroth hives, bees are portable. Nothing stops farmers coming to some financial arrangement with beekeepers to locate hives amid their crops. A couple of decades after James Meade's famous example, another economist, Steven Cheung, became curious about it, and he did something we economists perhaps don't do often enough: he called up some real people and asked them what actually happens.[7] It turned out that apple farmers pay beekeepers for the service of pollinating their crops. For some other crops, the beekeepers indeed pay farmers for the right to harvest their nectar, the market Meade said should exist but could not; one example is mint, which doesn't need any help from bees but which produces good honey.

So apples and bees aren't a good example of a positive externality, because the interaction *does* take place in a marketplace. And that marketplace is huge. These days, its centre of gravity is the California almond industry. Almonds occupy almost a million acres of California, and farmers sell $5 billion worth: that's just the price at the farm gate.[8] Almonds need honeybees – five colonies per hectare, rented for around $185 a colony.[9] Langstroth hives are duly strapped together, loaded onto the back of articulated lorries, 400 hives per truck, and driven to the Californian almond groves each spring, travelling by night while the bees are asleep.

The numbers are astonishing: 85 per cent of the 2 million commercial hives in the US are moved, containing tens of billions of bees.[10] The big beekeepers manage 10,000 hives each,[11] and from California they may travel up to Washington state's cherry orchards, then east to the sunflowers of North and South Dakota, and then on to the pumpkin fields of Pennsylvania, or the blueberries of Maine.[12] Meade was quite wrong to imagine beekeeping as some kind of rural idyll. Bees have been almost fully industrialised and pollination thoroughly commercialised.

And that presents a conundrum. Ecologists are worried about wild bee populations, which are in sharp decline in many parts of the world. Nobody quite knows why: candidates for blame include parasites, pesticides and the mysterious 'colony collapse disorder' where bees simply disappear, leaving a lone queen behind. Domesticated bees face the same pressures, so you might expect to see some simple economics at work – a reduction in the supply of bees increasing the price of pollination services.

But that's not what economists see at all. Colony collapse disorder appears to have had minimal effect on any practical metric in the bee market: farmers are paying much the same for pollination; the price of new queens, which are bred specially, has hardly budged. It appears that industrial beekeepers have managed to develop strategies for maintaining the populations on which they rely: breeding and trading queens, splitting colonies and buying booster packs of bees. That is why there is no shortage of honey – or almonds, or apples, or blueberries. Not yet, anyway.[13]

Should we celebrate economic incentives for preserving at least some populations of bees? Well, maybe. Another perspective is that it's precisely the modern economy's long-standing drive to control and monetise the natural world that caused the problem in the first place. Before monocrop agriculture changed ecosystems, there was no need to lug Langstroth hives around the countryside to pollinate crops – local populations of wild insects did the job free of charge.

So if we want an example of a positive externality – something the free market won't provide as much of as society would like – perhaps we should look to land uses that help wild bees and other insects. Wildflower meadows, perhaps. And some governments are indeed subsidising them – just as James Meade would have advised.[14]

36

Dams

Not far from Cairo stands a remarkable dam, the Sadd el-Kafara. It is more than 100 metres long, and 14 metres high, made of tens of thousands of tons of rock and earth, and could store about half a million cubic metres of water. These statistics are modest, by modern standards – but the Sadd el-Kafara is not a modern dam. It is nearly five thousand years old.[1]

And it was a spectacular failure: archaeologists believe the dam burst almost at once. Its centre is completely destroyed, the result of flood water spilling over the top of the structure and rapidly scouring away the downstream face of the dam, which crumbled like a sandcastle. We don't know who'd ordered the dam to be built. We can imagine their popularity took a hit.

One cannot fault the ancient Egyptians for trying. Water was scarce, and rainfall uneven. A sudden storm would have delivered a valuable resource, free of charge, literally falling from the sky – which would then have drained away towards the Mediterranean. A dam would have allowed water to be stored until needed.

Ancient Egypt isn't the only place to have found itself trying to deal with uneven rainfall. Much of the world's population lives in places where the availability of water is seasonal, or – increasingly – unpredictable. The plentiful, year-round water we have come to expect in developed countries often relies on a system of dams and reservoirs.

Where such a system is lacking, the effects can be brutal: Kenya lost more than 10 per cent of its economic output to drought in the late 1990s, followed by an even larger economic loss because of flooding.[2] With dams offering the potential to manage both droughts and floods, small wonder they have been tempting projects for millennia.

As a bonus, dams can include hydroelectric power stations, taking advantage of the gravitational potential of the pent-up water to turn turbines and produce clean electricity. Hydroelectricity is a bigger energy source than nuclear, solar, wind or tidal – and in many places, bigger than all three combined.[3] What's not to like?

The citizens of Henan province in central China could tell you. They live downstream of the Banqiao dam, which was built in the 1950s, and immediately showed signs of cracking. After reinforcement it was dubbed 'the Iron Dam' and declared unbreakable. In August 1975, it broke. Locals describe the event as 'the coming of the river dragon'. It was a wave several metres high and, eventually, 12 kilometres wide. Tens of thousands of people died – some estimates suggest almost a quarter of a million.

The tragedy was a state secret in China for many years.[4] It is the kind of catastrophe that puts Chernobyl into a new perspective. Nevertheless, it did not stop the Chinese government from deciding to replace the dam. (The Banqiao disaster was notable only because of its size; three thousand other dams failed in communist China between 1949 and 1980.)[5]

Even in wealthy countries, dams have been responsible for some of the deadliest man-made disasters. Large reservoirs weigh over 100,000 million tonnes when full – enough to cause earthquakes – and much smaller ones can still cause deadly landslides.[6] The Malpasset dam in France cracked in 1959 when the land at one edge of the curved concrete shell slipped under the pressure of the water. A total of 423 people died. Four years later, the new Vaiont dam in Italy was overwhelmed by an inland tsunami when the weight of its slowly filling reservoir caused a landslide. Nearly 2000 people died.[7]

Dams were military targets in the Second World War and the Korean war – but given the risk to civilians, military attacks on dams are now regarded as war crimes, and with good reason.[8] And a dam does not have to be destroyed to be a weapon. The Itaipu dam, on the border of Brazil and Paraguay, lies upstream of Argentina's capital, Buenos Aires. If the sluice gates were all opened at once, the city would be flooded.[9]

Yet it is not the risk of catastrophe that gives the modern dam its uneasy reputation: it is the harm done as the dam reshapes the ecosystem both up- and downstream.

The poster child for this harm has long been the High Aswan dam in Egypt. The High Aswan holds back the river Nile, creating a reservoir 500 kilometres long. *The Economist* gives a checklist of the consequences: 'an explosion of water hyacinth, outbreaks of bilharzia, polluted irrigation channels and a build-up of sediment inland that would otherwise compensate for coastal erosion from Egypt to Lebanon'.[10]

That doesn't even mention the fact that ancient Nubian temples were flooded, relocated to high ground, or moved in their entirety to museums in places such as Madrid, New York and Turin. And it wasn't just the temples that were displaced: more than 100,000 people were forced to move.[11]

But some experts argue that despite all the costs, the project has been an overwhelming success. The High Aswan allows predictable irrigation in both Egypt and Sudan. That is no small thing. It paid for itself within two years and shielded Egypt from what would have been a disastrous drought throughout the 1980s, followed by potentially catastrophic floods in 1988.

All dams create winners and losers – and tensions that need to be managed. The only two women to win the Nobel memorial prize in economics both studied dams. Elinor Ostrom showed how dams in Nepal destabilised traditional bargains between upstream and downstream communities about sharing water and sharing effort.[12] Esther Duflo found that large dams in India benefited some communities through irrigation, but increased poverty in others.[13]

The losers from dams often live in other countries, which makes the tensions international: nearly half the world's land area drains into rivers that cross international borders.[14] The latest example is the Ethiopian Renaissance dam, which will be Africa's largest hydroelectric project when completed in 2022. It is upstream from the High Aswan and has the potential to restrict the flow of the Nile towards Egypt. Egypt is not happy.[15]

But compensating the losers is not always a priority for politicians. They're often more interested in the symbolism. You can see why. Collapsed dams like the Sadd el-Kafara and Banqiao speak of terrible errors of judgement. From the early Soviet Union's Dneprostroi to modern China's Three Gorges dam – which vies with the Itaipu for the title of the world's largest power station – political leaders have wanted successful dams to stand testament to their grand strategic vision.

Some believe the enduring bad reputation of the High Aswan dates from Cold War propaganda. When Egypt's

President Nasser couldn't get backing for the dam from the United States, he turned to the Soviet Union, and national-ised the Suez Canal to help pay for it. That led to the Suez crisis. No wonder Western leaders didn't want Nasser to get a public-relations boost.[16]

Dams reshape economies in complex ways. Many may be worth it overall only if the benefits can be equitably shared with those who lose out. But this messy reality is easy to overlook when dams are seen as symbols of national virility. When India's first Prime Minister, Jawaharlal Nehru, spoke to villages displaced by the colossal Hirakud dam project in 1948, he may have been more plain than he intended: 'If you are to suffer, you should suffer in the interest of the country.'

It is not clear that anyone found that comforting.[17]

VII

NO PLANET B

37

Fire

'The canyons seemed to act as chimneys, through which the wind and fires swept with the roar of a thousand freight trains. The smoke and heat became so intense that it was difficult to breathe ... The whole world seemed to us men back in those mountains to be aflame. Many thought that it really was the end of the world.'[1]

It was 20 August 1910, and Forest Ranger Ed Pulaski was caught in the middle of what would become known as the 'Big Blowup'. Pulaski realised that his task was no longer to save the forests of northern Idaho, but to save the firefighters. He charged around on his horse, eventually gathering 45 men.

Trees were falling all about us under the strain of the fires and heavy winds, and it was almost impossible to see through the smoky darkness. Had it not been for my familiarity with the mountain trails, we would never have come out alive, for we were completely surrounded by raging, whipping fire. My one hope was to reach an old mine tunnel which I knew to be not far from us. We raced for it. On the way one man was killed by a falling tree. We reached the mine just in time, for we were hardly in when the fire swept over our trail.[2]

Pulaski passed out. The next morning, he couldn't see. His hands were burned. But he was alive, and so were all but five of his men. The Big Blowup had killed 86 people, and consumed enough wood to build 800,000 houses.[3] It also seared the national consciousness: the US Forest Service promised to douse all wildfires as quickly as they could.[4]

That was unwise – we'll come back to why – but you can understand it: fire is terrifying. It's also fundamental to the modern economy. And its story goes back much further.

Still, for the first 90 per cent of Earth's history, there was no fire at all. There were volcanic eruptions – but molten rock isn't *on fire*, because fire is a chemical reaction: the process of combustion.[5] It's life that creates both the oxygen and the fuel that fires need to burn. Fossil evidence suggests that flammable plant life evolved around 400 million years ago, and periodically went up in smoke, due partly to those volcanoes but mostly to lightning. In recent years satellite observations have shown us how surprisingly common lightning is – there are around 8 million strikes a day. It's still responsible for more wildfires than ill-advised barbecues or carelessly discarded cigarette butts.[6]

Fire shaped landscapes – and, with it, evolution. It enabled the spread of grasslands, somewhere around 30 million years ago; without fire, they'd have reverted to scrub or forest. And grasslands are thought to have played a role in the emergence of the hominins who evolved into us.[7]

Try to imagine the economy before our ancestors tamed fire. You can start by discarding any products made with metal, or using metal tools – metal starts life in a furnace. The same goes for glass. Now forget anything that involves burning fossil fuels, for transport or electricity; or that uses materials you need fire to make heat to produce – think plastics, or plants grown with artificial fertilisers, made with the Haber-Bosch process. No bricks or pottery: they're fired in a kiln. There isn't much left. Raw, organic food, cut up with a sharp stone? We can hardly call it an 'economy' at all.

Exactly when and how our ancestors learned to control fire is a matter of some debate, but it's unlikely to have played out as famously imagined by Disney's *Jungle Book*, with ape King Louis begging Mowgli for the secret of 'man's red fire'.[8] In fact, chimps appear to understand pretty well how wildfire will spread.[9] And other species are reportedly alert to the hunting opportunities.[10] Some birds of prey have even been seen picking up burning sticks, dropping them to start a new fire, and pouncing on the creatures who then make a run for it.[11]

It seems likely that our ancestors similarly harnessed wild-fires for hundreds of thousands of years before they figured out how to make sparks from flint.[12] Perhaps they kept the fires alive by adding slow-burning animal dung.[13] No doubt they'll have used them for hunting, keeping warm, and fending off predators.[14] They'll have cooked with fire; the primatologist Richard Wrangham argues that as cooked food provides more energy, it enabled humans to evolve bigger brains.[15] Meanwhile, the archaeologist John Gowlett links fire to the 'social brain' hypothesis – the idea that we evolved bigger brains to navigate growing social pressures; evenings around the fire will have given our ancestors more time to socialise.[16]

However much truth is in those speculations, economic development has seen us confine fire to various special chambers – from industrial plants to internal combustion engines to the gas oven in your kitchen. The historian Stephen Pyne calls this the 'pyric transition'.[17] And where that's not yet happened, it's a problem: in developing countries, millions of deaths are linked to air pollution caused by cooking on indoor fires.[18] But Pyne argues that this transition increased our fear of wildfires. And with climate change we can expect to see more of those fires. While satellite observations are helping us to understand wildfires, changing patterns of weather and vegetation are making them harder to predict.[19]

It took half a century after Ed Pulaski's heroics for consensus to form that quickly extinguishing wildfires isn't such a great idea. The problem is that eventually there'll be a fire you can't control – and that fire will be more devastating, because it will burn through all the deadwood that would have been cleared by the small fires if you hadn't rushed to put them out.

And in the meantime, complacency sets in: we're increasingly building in or close to wilderness areas where fires will break out sooner or later, from California to Australia. When experts advise it might be wise to let those fires burn, you can bet that the people who live nearby aren't going to be too keen.[20] As Andrew Scott argues in his book *Burning Planet*, 'Our increasing scientific understanding of fire in recent years has not translated into greater public awareness.'

Some economists think that wildfires are just one example of a broader modern dynamic: that getting better at handling small problems creates a growing sense of safety, which paradoxically creates the risk of much larger problems. Greg Ip has applied this analysis to the financial crisis of 2007–08. Policymakers had got so good at extinguishing minor crises that people became overconfident, and took silly risks – such as betting the ranch on subprime mortgages. And when a crisis came along that couldn't be stamped out, those bad bets fuelled a global conflagration.[21]

38

Oil

The message had been sent. Edwin Drake's last financial backer had finally lost patience. *Pay off your debts,* read the message, *give up, and come home.*[1]

Drake had been hoping to find 'rock oil', a brownish crude that sometimes bubbled near the surface of western Pennsylvania. He planned to refine it into kerosene, for lamps – a substitute for the increasingly expensive whale oil. There would also be less useful by-products, such as gasoline, but if he couldn't find a buyer for that he could always pour it away.

The message had been sent, but Drake had not yet received it when his drill bit punctured an underground reservoir full of crude oil under pressure. From 69 feet under the surface, the oil began to rise. It was 27 August 1859. The whales had been saved, and the world was about to change.

Just a few miles south and a few years later came a hint of what lay in store. In 1864, when oil was struck at Pithole, Pennsylvania, 'there were not 50 inhabitants within half a dozen miles' according to the *New York Times*. A year later, Pithole had at least 10,000 inhabitants, fifty hotels, one of the country's busiest post offices, two telegraph offices, and dozens of brothels.[2]

A few men made fortunes, but a real economy is complex and self-sustaining. Pithole was neither. Within another year it was gone, its wooden buildings burnt down or dismantled and moved ten miles to the next strike, in what became the imaginatively named Oil City.[3]

Pithole's oil boom did not last, but our thirst for the fuel grew and grew. The modern economy is drenched in oil. It's the source of more than a third of the world's energy. That's more than coal; it's also more than twice as much as nuclear, hydroelectric and renewable energy sources combined. Oil and gas together provide a quarter of our electricity.[4] And the raw material for most plastics.

Then there's transport. Edwin Drake may have wondered who would buy gasoline, but the internal combustion engine was about to solve that problem. By 1904, Standard Oil controlled more than 90 per cent of US oil refining – and oil was so important that the US government decided to break it up, into companies that eventually became the likes of Exxon, Mobil, Chevron and Amoco.[5] From cars to trucks, cargo ships to jet planes, oil-derived fuel still moves us – and stuff – around.[6]

No wonder the price of oil is arguably the most important single price in the world. In 1973, when some Arab states declared an embargo on sales to several rich nations, prices surged from $3 to $12 a barrel in just six months. A global recession followed. It wasn't the last: US recessions followed price spikes in 1978, 1990 and 2001. Some economists even believe that record-high oil prices played an important role in the global recession of 2008, which is conventionally blamed on the banking crisis alone. As oil goes, so goes the economy.[7]

So why did we become so excruciatingly dependent?

Daniel Yergin's magisterial history of oil, *The Prize*, begins
with a dilemma for Winston Churchill. Made head of the
Royal Navy in 1911, Churchill had to decide whether the
British empire would meet the challenge of an expansionist
Germany with new battleships powered by safe, secure Welsh
coal, or by oil from faraway Persia – modern-day Iran. Why
would anyone rely on such an insecure source? Because
oil-fired battleships would accelerate faster, sustain a higher
speed, required fewer men to deal with the fuel, and would
have more capacity for guns and ammunition. Oil was simply
a better fuel than coal. Churchill's 'fateful plunge' in April
1912 reflected the same logic that has governed our depend-
ence on oil – and shaped global politics – ever since.[8]

After Churchill's decision, the British Treasury bought a
majority stake in the Anglo-Persian oil company – the ances-
tor of BP. In 1951, the government of Iran nationalised it.[9]
Our company, protested the British. Our oil, responded the
Iranians. The argument would be repeated around the world
over the subsequent decades.

Some countries did well. Saudi Arabia is one of the richest
on the planet. Its state-owned oil company, Aramco, is worth
more than Apple or Google or Amazon.[10] Still, nobody would
mistake Saudi Arabia for a complex, sophisticated economy
such as that of Japan or Germany.* It's more like Pithole on
a grander scale. Elsewhere, from Iraq to Iran, Venezuela to
Nigeria, few oil-rich countries have prospered from the discovery. Economists call it 'the curse of oil'.[11]

Juan Pablo Pérez Alfonzo, Venezuela's oil minister in the
early 1960s, had a more vivid description. 'It is the devil's

---

* MIT's 'Observatory of Economic Complexity' ranks Saudi Arabia as the 32nd
most complex economy in the world. It is the 14th richest, measured by per capita
income. Qatar is a more extreme example: the richest country in the world, but
only the 66th most complex.

excrement,' he declared in 1975. 'We are drowning in the devil's excrement.'[12]

Why is it a problem to have lots of oil? When you export it, that pushes up the value of your currency – which makes everything other than oil cheap to import, and prohibitively expensive to produce at home. That means it's hard to develop other economic sectors, such as manufacturing or complex services. Meanwhile politicians often focus on monopolising the oil, for themselves and their allies. Dictatorships are not uncommon. There is money – for some – but such economies are thin and brittle. At least in Pithole, people could leave when the well ran dry. It's not so easy to stroll away from an entire country.

That's one reason we might hope for something to replace oil. Climate change, obviously, is another. But oil has so far stubbornly resisted giving way to batteries. This is because machines that move around need to carry their own source of power with them, the lighter the better. A kilogram of petrol stores as much energy as 60 kilograms of batteries,[13] and has the convenient property of disappearing after use. Empty batteries, alas, are just as heavy as full ones. Electric cars are finally starting to break through. Electric jumbo jets are a tougher challenge.

There was a time when it seemed as though oil might simply start to run out – 'peak oil' was the phrase – pushing prices ever higher and giving us the impetus to move to a clean, renewable economy. Now it looks as though we will have to be more determined if that's what we really want. The oil market has been transformed by the rapid growth of hydraulic fracturing, or 'fracking', in which water, sand and chemicals are pumped underground under high pressure to crack open rocks and release oil and gas. Fracking is more like manufacturing than traditional exploration and production:

it's standardised, enjoying rapid productivity gains, and the process starts and stops depending on whether the price is right. Between 1980 and 2015, oil was discovered twice as quickly as it was consumed.[14]

We are still drowning in the devil's excrement, it seems, and it is only going to get deeper.

39

Vulcanisation

The black-and-white photograph shows a man, perched on the edge of a wooden deck, looking down at two objects. At first, you can't take in what they are. In the background are palm trees; two other men, one with arms folded, one with hands on hips, look grimly at their friend – or perhaps at the photographer, it's hard to tell.

The photographer was Alice Seeley Harris; the year 1904; the location a missionary outpost in Baringa, in what was then called the Congo Free State. The man's name was Nsala. As he explained to Alice, his wife and children had just been killed. And here, he told her, opening up a bundle of leaves, was the proof – all that the attackers had left of his five-year-old daughter, Boali.

Alice's photograph of Nsala, looking at his daughter's severed hand and foot, caused an uproar back in Europe. And she found no shortage of other brutal sights at which to point her Kodak camera.[1] Children with hands cut off. Women chained at the neck. Vicious hippo-skin whips called the *chicotte*; beatings were commonly fatal.[2]

Printed in pamphlets and displayed at public meetings,

Alice's harrowing images formed the world's first photo-graphic human-rights campaign.[3] They helped build public pressure that eventually forced Belgium's King Leopold to loosen his grip on the colony famously depicted in Joseph Conrad's novel *Heart of Darkness*. As the main character, Kurtz, exclaimed, 'The horror!'[4]

But why was Leopold's Congo so horrific?

Rewind seven decades: New York, 1834. An impover-ished, unwell, but preternaturally optimistic young man is knocking on the door of the Roxbury India Rubber Company. Charles Goodyear had landed in debtors' prison when his family's hardware business went bankrupt, but he was sure he could invent his way out of financial trouble. His latest idea: an improved kind of air valve for inflatable rubber life preservers.

Goodyear was in for a surprise. The manager loved his valve – but confessed that his company was on the verge of ruin. He rued the day he'd got into the rubber business.[5]

He wasn't alone. All over the country, investors had sunk money into this miraculous new substance – stretchy and pliable, airtight and waterproof – and now it was all going horribly wrong.

Actually, rubber wasn't exactly new. It had long been known to South Americans, and first reported by Europeans in the 1490s: the natives made 'a kind of wax' from trees that 'give milk when cut'.[6] That 'milk' was latex – it comes from between the inner and outer bark.

Bits of rubber made their way to Europe, but were mostly a curiosity. In the 1700s a French explorer brought home the name *caoutchouc* from a local language; it meant 'weeping wood'. But it was the scientist Joseph Priestley who bestowed the name we now commonly use, when he noticed it rubbed pencil marks off paper.

By the 1820s, rubber was attracting serious interest. More and more was being shipped from Brazil, and made into coats, hats, shoes, those inflatable life preservers. Then came a really hot summer, and entrepreneurs watched aghast as their inventories simply melted into foul-smelling goo.[7]

Goodyear saw his chance. A fortune awaited whoever could invent a way to make rubber cope with heat – and cold, which made it brittle. And he was the man to do it. True, he had no background in chemistry, and no money, but why should that stop him?

For years Goodyear dragged his wife Clarissa and their growing brood from town to town, renting ever more insalubrious houses, pawning their dwindling stock of possessions, running up debts and intermittently being hauled off to debtors' prison, but always just about managing to find some new relative to sponge off or investor to convince that a breakthrough was just around the corner. When Clarissa wasn't trying to feed the children, Charles commandeered her saucepans to mix rubber with – well, anything he could think of: magnesium, lime, carbon black. Once he filled the kitchen with nitric acid fumes and was bed-ridden for weeks.[8]

In the end, he found the answer: heat the rubber with sulphur. It's a process we now call vulcanisation. Sadly for the long-suffering Clarissa, it led to her husband borrowing still more money for lawsuits to try to protect his patents: when he died, he owed $200,000. But Charles's doggedness had put rubber at the very heart of the industrial economy: in belts and hoses and gaskets; sealing, insulating, absorbing shocks.[9]

In the late 1880s, a Scotsman living in Ireland supplied the killer app: the pneumatic tyre. John Boyd Dunlop was a veterinary surgeon; he'd been tinkering with his son's tricycle, trying to find a way to cushion the ride. Bike manufacturers quickly saw the advantages; so did the nascent car industry.

Demand for rubber boomed. Europe's colonial powers set about clearing vast areas of Asia to plant *Hevea brasiliensis* – more widely known as the 'rubber tree'.[10]

But those new rubber-tree plantations would take time to grow, and hundreds of other plants also produce latex, in varying quantities – even the humble dandelion.[11] In the Congo's rainforest were vines that could be tapped to meet demand right now.[12]

How to get that rubber, as much and as quickly as possible? In the absence of scruples, the answer was distressingly simple: send armed men to a village; kidnap the women and children; and if their menfolk didn't bring back enough rubber, chop off a hand – or kill a family.[13]

Some things have changed since Nsala met Alice Seeley Harris in Baringa. More than half the world's rubber now comes not from weeping wood but gushing oil.[14] Attempts to make synthetic rubber began as the natural stuff grew popular, and took off during the Second World War. With supply lines from Asia disrupted, America's government pushed industry to develop substitutes.[15] Synthetic rubber is often cheaper, and sometimes better – including for bicycle tyres.[16]

But for some uses, you still can't beat a bit of *Hevea brasiliensis*.[17] About three-quarters of the global rubber harvest goes into tyres for heavier vehicles.[18] And as we make more cars and trucks and planes, we need ever more rubber to clothe their wheels. That's not without problems. The rubber tree is thirsty: environmentalists worry about water shortages – and biodiversity, as south-east Asia's tropical rainforest increasingly gives way to plantations.

It's happening in Africa, too. Travel a thousand kilometres through the rainforest from Baringa, for example, bearing west and slightly north, and you'll come to Meyomessala in Cameroon. Nearby, the world's largest rubber-processing company – majority-owned by the Chinese state – is clearing thousands of hectares for rubber trees. The company says it's committed to ethical sourcing; villagers say they haven't been properly compensated for loss of their lands.[19]

So rubber is still causing controversy, but now it's cutting down trees, not cutting off hands. It is progress, of a sort.

40

The Wardian Case

Robert Fortune was 'much annoyed' when his Chinese servant returned to the ship with a pathetically small collection of plants – clearly, rather than trekking into the hills, the servant had barely ventured beyond the shore. Fortune assumed he'd been lazy: 'like most of the Chinese . . . he was rather remarkable for this propensity.' The servant protested: he'd been told that the people who lived in the hills in that part of south-east China were dangerous. Nonsense, said Fortune. They'd go together.

The ship's captain offered to send some crew members for protection. No need, said Fortune. Doubts started to creep in when the locals, seeing him stride off towards the hills, 'attempted to dissuade me from going, by intimating that I was sure to be attacked by the Chinchew men, and robbed or murdered'. Then he noticed the locals were armed; self-defence, explained his servant. Well, it was too late now: 'I determined to put a bold face on the matter and proceed.'

At first, all went smoothly. Foreigners were a rare sight, and Fortune drew quite a crowd, but they were 'generally civil'; he filled his specimen boxes, with 'three or four hundred of the Chinese, of both sexes, and all ages, looking down upon us with wonder'. Soon, however, Fortune had been deftly pickpocketed and his servant – 'pale with fright' – was surrounded by knife-wielding robbers: 'My poor plants collected with so much care were flying about in all directions.'

This episode doesn't appear to have affected the young Scotsman's confidence. Later in the trip, he's warned that an area he wants to sail to is swarming with pirates: '"Nonsense!" I exclaimed: "no pirates will attack us."' You can guess what happened.

But Fortune made it back to Shanghai, where he 'despatched eight glazed cases of living plants for England'. He concludes his 400-page memoir of the expedition by noting with satisfaction that 'the *Anemone japonica* is in full bloom in the garden of the Society at Chiswick'.[1]

Fortune was a plant-hunter, employed by those people in Chiswick – the Horticultural Society of London, now the Royal Horticultural Society. And those glazed cases were making it a whole lot easier to plant-hunt. They were called Wardian cases, and they'd been developed in the previous decade, the 1830s. Nathaniel Bagshaw Ward was a doctor in London's East End. He was also a fern enthusiast, but he struggled to grow ferns because the city air was so polluted.[2]

Ward's invention was simple, and in retrospect obvious. Glass, timber, putty, paint – it was essentially a sealed mini-greenhouse. It let the light in. It kept the soot and smoke out. And it kept the moisture in, so there was no need to water the plants. This was no feat of technology, but the result of a questioning mind. It was commonly assumed that plants needed the open air. Ward wondered: what if they didn't?

His ferns thrived. And Ward soon realised he might have solved a problem that vexed plant-hunters – how to keep their plants alive on a long sea journey. Put them below deck, and they suffered from lack of light. Put them above deck, and they suffered from salty spray.[3] And if water ran low, the crew would probably rather the plants went thirsty than they did.[4]

Ward arranged an experiment: he shipped two cases of plants to Australia. Several months later came a letter from the ship's captain, offering 'warm congratulations': most ferns were 'alive and vigorous', and the grasses were 'attempting to push the top of the box off'. The ship returned with Ward's cases packed with Australian plants – again, perfectly healthy.[5]

Ward wrote a book: *On the Growth of Plants in Closely Glazed Cases*. He was sure his invention would have far-reaching impacts, and he was right – but not in the way he expected. Ward imagined that as the 'higher and middle classes' bought his cases to cultivate plants at home, it would create 'a new field of healthful industry' for the poor – procuring plants from the countryside. He thought humans, like ferns, would benefit from being out of the polluted London air, and he envisaged large sealed greenhouses in which people could convalesce from measles or consumption.

He didn't foresee that his case was about to reshape global agriculture, politics and trade. Perhaps he should have, because plant hunting had never been only about herbaceous perennials. The 'father of modern plant hunting', Sir Joseph Banks, was keenly aware of the economic potential of moving crops from one colonial outpost to another.[6]

In the late 1700s he turned London's Kew Gardens into a kind of imperial clearing house for flora. It was Banks who caused Captain William Bligh to embark on his ill-fated voyage on the HMS *Bounty*, a voyage that was to end in an infamous mutiny. Bligh was supposed to deliver breadfruit plants to the West Indies. Banks hoped they'd become a cheap way of feeding slaves.[7]

Thanks to the Wardian case, the process of transplanting –

well, plants – now had wind in its sails. Before, said one commercial importer, they expected 19 out of 20 plants to die at sea. In Ward's cases, they found, 19 out of 20 were surviving.[8]

It was Wardian cases, for example, that spread the Cavendish banana around the world. That's the variety you see in shops today. William Cavendish was the president of the Horticultural Society.[9]

Wardian cases destroyed Brazil's rubber industry. With prices high, the British Foreign Office sent an enterprising amateur botanist to the Amazon to sneak out some rubber seeds. They germinated in Kew, and seedlings were shipped to East Asia. Brazil couldn't compete with colonial plantations.[10]

And Wardian cases helped to break China's grip on the tea market. Ward published his book in the year Britain won the First Opium War: when the Chinese decided to stop accepting Indian-grown opium in exchange for their tea, the British sent in gunboats to change their minds. You can see why: taxes on tea accounted for nearly a tenth of the British government's income.[11]

Now the East India Company, which virtually governed the subcontinent on Britain's behalf, decided they needed a backup approach: grow more tea in India. As Sarah Rose writes in For All the Tea in China, 'The Indian Himalaya mountain range resembled China's best tea-growing regions.'[12]

That meant they needed to smuggle tea plants out of China. And there was only one man for that job. Robert Fortune had learned on his first expedition that if he shaved his head and wore a wig and Chinese clothes, he could just about pass unnoticed: 'upon the whole,' he wrote, 'I believe I made a pretty fair Chinaman'.[13] Suitably disguised, he eventually shipped nearly twenty thousand tea plants for his new employers.[14]

But perhaps the most significant impact of the Wardian case wasn't bringing plants to Europe from more far-flung places – it was enabling more people from Europe to go to far-flung places. Wardian cases allowed the cinchona tree to be shipped from South America to India and Sri Lanka.[15] From its bark came quinine, which helped ward off malaria. That made it less scary for Europeans to venture to the tropics; some historians think Africa might not have been colonised without it.[16] After all, not every traveller was as blithely impervious to risk as Robert Fortune.

41

Cellophane

You're the top! You're Mahatma Gandhi.
You're the top! You're Napoleon brandy.

This is a Cole Porter song, written in 1934. And what else might he compare the object of his affections to? A summer's day? Nope.

You're the purple light of a summer night in Spain . . .
You're cellophane.[1]

The latest in transparent food packaging. Of course! That wouldn't happen nowadays, and not just because the less mellifluous 'you're low-density polyethylene' doesn't rhyme.

Plastic packaging has a bad (w)rap. When the UK's *Guardian* newspaper invited readers to share examples of annoyingly unnecessary packaging, comments flooded in: shrink-wrapped cucumbers; apples in hard plastic tubes; pre-cut melon in little pouches; bananas in bags.[2] Doesn't Mother Nature already provide bananas with packaging of their own? It all seems so obviously wasteful.

We'll come back to that seeming obviousness. But let's start our packaging story in a more innocent age – before anyone worried about plastic in landfills, or the sea, or the food chain.[3] It begins in 1904, at an upmarket restaurant in the Vosges, in France, when an elderly patron spills red wine over a pristine linen tablecloth. Sitting at a nearby table is a Swiss chemist called Jacques Brandenberger. He works for a French textile company, and as he watches the waiter change the tablecloth, he finds himself wondering: could he make a fabric that would simply wipe clean?[4]

He couldn't: he tried spraying cellulose on tablecloths, but it peeled off in transparent sheets. But might those transparent sheets have a market? By the First World War, he'd found one: eyepieces for gas masks. He called his invention cellophane, and in 1923 he sold the rights to the DuPont corporation in America.[5] Early uses there included wrapping chocolates, perfume and flowers;[6] perhaps those romantic connotations inspired Cole Porter.

But DuPont had a problem. Some customers weren't happy. They'd been told cellophane was waterproof, and it was, but it wasn't *moisture-proof*. Candies were sticking to it; knives were rusting in it; cigars were drying out.[7] DuPont hired a 27-year-old chemist, William Hale Charch, and tasked him with finding a solution. Within a year, he'd figured one out: coating the cellophane with extremely thin layers of nitro-cellulose, wax, a plasticiser and a blending agent.[8]

Sales took off. The timing was perfect: in the 1930s, super-markets were changing – customers no longer queued to tell shop assistants what food they required; they picked products off the shelves instead. See-through packaging was a hit.[9] One study found that cellophane-wrapping crackers boosted sales by more than half.[10] Admittedly that study was conducted by DuPont, but retailers had no shortage of similar advice. A less-than-progressively-titled article in *The Progressive Grocer* announced: 'she Buys Meat with Her Eyes.'[11]

In fact, the meat counter was the hardest to make self-service. The problem was that meat, once cut, would quickly discolour. Trials suggested that by sparing shoppers from the queue to instruct the butcher, self-service could sell 30 per cent more meat – and with such an incentive, solutions would be found: pink-tinted lighting; antioxidant additives; and – of course – an improved version of cellophane, that let through just the right amount of oxygen. By 1949, DuPont adverts read: 'The pleasing new way to buy meats – self-service, just like other foods. Pre-cut, weighed, priced and wrapped in Cellophane right in the store'.[12]

But cellophane would soon fall out of fashion, outcompeted by the likes of Dow Chemical's polyvinylidene chloride. Like cellophane, it was an accidental discovery that was first used in conflict – in this case, weatherproofing fighter planes in the Second World War. And, like cellophane, it needed plenty of research and development before it could be used on food: it was originally dark green and smelled disgusting. Once Dow sorted that out, it hit the market as Saran wrap – now more widely known as cling wrap, or clingfilm.[13]

After health scares with polyvinylidene chloride, clingfilm is now often made with low-density polyethylene, though that's less – well, clingy.[14] Low-density polyethylene is also the stuff of those one-use supermarket bags that are now being banned around the world.[15] 'And what's *high-density* polyethylene?', I hear you ask. That's what you might get milk in. Not fizzy drinks, though; they come in polyethylene terephthalate.[16] And if you're not lost already, consider that plastic packaging is increasingly made from multiple layers of these and other substances, such as biaxially oriented polypropylene or ethylene–vinyl acetate.[17]

There's a reason for this, say packaging gurus: different materials have different properties, so multiple layers can give you the same performance from a thinner – and thus lighter – piece of packaging. But these compound packaging materials are harder to recycle. The trade-off is hard to fathom. Depending on how much of the heavier, recyclable packaging would in practice be recycled, you might find that the lighter, non-recyclable packaging actually generates less trash.[18]

And once you start looking into plastic packaging, this kind of counter-intuitive conclusion comes up all the time. Some packaging is a foolish waste. But are shrink-wrapped cucumbers really so silly? They stay fresh for 14 days, against just three days without the wrapping.[19] Which is worse: the gram-and-a-half of plastic wrap, or the waste of cucumbers going off before being eaten? Suddenly it's not so obvious.

Plastic bags stop bananas going brown so quickly, or new potatoes going green; they catch grapes that fall off bunches.[20] About a decade ago, a UK supermarket experimented with taking all its fruit and vegetables out of their packaging, and their wastage rate doubled.[21] The last thing we want is super-markets using plastics behind the scenes, then stripping away the packaging before we see it. All the same plastic waste, doing less good.

And it's not just shelf life – there's waste en route to the shelves. Another supermarket, stung by criticism for putting apples in plastic-wrapped trays, tried selling them loose from big cardboard boxes – but so many were damaged in transit, this used *more* packaging per apple actually sold.[22] According to a UK government report, only 3 per cent of food is wasted before it gets to stores – in developing countries that figure can be 50 per cent, and that difference is partly due to how the food is packaged.[23] As more of us live in cities, far from where food is grown, this matters.

Even the dreaded single-use shopping bag might not be the villain it seems. If you've bought sturdy, reusable bags from your supermarket, it's likely they're made from non-woven polypropylene – and they are less damaging, but only if you use them at least once a week for a year. That's according to a report by the Danish government, which weighed up the varied environmental impacts of producing and disposing of different kinds of bag.[24] If your reusable bag is organic cotton, don't feel smug: the researchers reckon they need twenty thousand uses to justify themselves.[25] That's a shopping trip every day, for over half a century.

The market can be a wonderful way of signalling popular desires. Shoppers in 1940s America wanted convenient, pre-cut meat – and the invisible hand delivered the technologies that made it possible. But our desire for less waste may not yield to market forces, because the issue is complicated and our choices at the checkout may accidentally do more harm than good. We can only send that message on a more circuitous route, through governments and pressure groups, and hope that they – and well-meaning industry initiatives – will figure out some sensible answers.[26]

One thing is clear: the answer won't be no packaging – it'll be better packaging, dreamed up in research and development labs of the kind that gave us moisture-proof cellophane. Maybe Cole Porter was onto something after all.

42

Recycling

Sail up the Pearl River estuary from Hong Kong, past Shenzhen, and you come to the industrial city of Dongguan. Here you'll find what may be the world's biggest paper mill, larger than 300 football pitches.[1] It's owned by Nine Dragons, a recycling company started by Zhang Yin, who was once ranked by *Forbes* as the world's richest self-made woman.[2]

Nine Dragons is – or, perhaps, was – the largest importer by volume of American goods into China.[3] Those goods? Waste paper – typically with some less useful trash mixed in. It's among many Chinese companies that built business models around importing what Americans and others put in their recycling receptacles, and picking out stuff that shouldn't be there.[4] That's a crucial job: if the waste is too contaminated, you can't recycle it.

It's also a job that's hard to automate. Perhaps one day robots will sift through our trash as deftly as Wall-E in the Pixar movie – but, for now, we need humans. So rich countries started shipping their waste to countries where workers are poor enough to sort it for wages low enough to turn a profit.[5]

From the 1980s until very recently, this system worked smoothly. China's fast-growing economy was exporting lots of manufactured goods. And because those ships would otherwise return empty, it was cheap to load them with waste for China to recycle.[6] Entrepreneurs such as Mrs Yin made a fortune.

But as China got richer, the government decided it no longer wanted to be a dumping ground for the rest of the world's badly sorted rubbish. It 2017 it announced a new policy, called National Sword, so strict that most experts thought it wouldn't happen. But on New Year's Eve that year it did: abruptly, China would now accept only well-sorted rubbish, no more than one half of 1 per cent stuff that shouldn't be there. That was a big change – contamination rates used to reach 40 times higher.[7] The amount of waste being shipped to China plunged.[8]

Governments and recycling companies scrambled to adjust. Should they find other countries still poor enough to accept their badly sorted waste? Or raise taxes to pay higher-wage workers to sort it better? Or something else?

Let's come back to that conundrum via a brief history of recycling. And here I should distinguish recycling from *reusing* – there's a reason the 'reduce, reuse, recycle' mantra is in that order. Take glass bottles: if you can rinse and refill them, that makes more sense than crushing and melting them to make new ones.

Examples of reuse go back before paper, to papyrus: Ancient Greece gave us the word 'palimpsest', which literally means 'scraped clean to be used again'.[9] As for recycling, the Romans melted old bronze statues to sculpt new ones.[10] A thousand years ago, Japan was pulping paper to make more paper.[11] For centuries people have scraped a living by scavenging for scrap, such as rags to sell to paper mills.[12] But that was all driven by saving or making money: the raw materials were too valuable to be thrown away. The idea that we should recycle because it's the right thing to do is much more recent.

To see how attitudes have changed, look at *Time* magazine from August 1955. There's an article headlined 'Throwaway Living', and the adjective isn't pejorative – it's celebratory. The subtitle is 'disposable items cut down household chores'. An image shows a smiling family filling their bin with paper plates, plastic cutlery and other objects that, the article tells us, 'would take 40 hours to clean – except that no housewife need bother'. Why wash up after cooking when you can use a foil 'Disposa-pan' – or a throwaway barbecue, complete with a handy asbestos stand?[13]

A TV ad campaign known as 'The Crying Indian' helped shift the mood, in America at least.[14] First shown in 1971, it depicted a Native American man paddling his canoe down a trash-polluted river; then he stands by a highway as a passing motorist tosses a bag of fast-food detritus at his feet. 'Some people have a deep, abiding respect for the natural beauty that was once this country. And some people don't. People start pollution. People can stop it.' The Native American man turns to the camera, a single tear rolling down his cheek.[15]

But the advert wasn't all it seemed, and not just because the actor later turned out to be a second-generation Italian immigrant.[16] It was funded by an organisation backed by leading companies in the beverage and packaging industries. At the time, deposit schemes were common: you'd buy a fizzy drink, and get some cash back when you returned the bottle. This model assumes it's the manufacturers' job to provide the incentives and logistics for returning waste.[17]

The Crying Indian had a different message. Who's responsible? *People*. Deposit schemes fell out of fashion. Recycling logistics became seen as a matter for local government.[18] The historian Finis Dunaway argues that turning 'big systemic problems into questions of individual responsibility' was a bad idea; it made recycling less about effective action, more about making ourselves feel good.[19]

That seems to chime with research by behavioural economists at Boston University, who found that when people know they'll be able to recycle, they act more wastefully.[20] That wouldn't matter if recycling was cost-free; of course, it isn't.

The economist Michael Munger develops a similar argument. You can't just leave waste disposal to the free market – if you charge people what it really costs, you tempt them to dump it illegally instead. You have to subsidise it. But that incentivises the behaviour in *Time* magazine – people chuck stuff away, and society bears the costs. How do we get them to recycle instead? One solution is moral suasion, for example The Crying Indian.

But that creates a problem, says Munger, in an essay for the libertarian think tank the Cato Institute. For each kind of waste – glass bottles, plastic coffee cups – what we should do is coolly compare the costs and benefits of recycling against other options. Well-designed landfills are nowadays pretty safe, and we can harness the methane they produce for electricity.[21] Modern waste incinerators can be a clean-ish source of power.[22] If instead we turn recycling into a moral good, when do we stop?

Which brings us back to the conundrum posed by China's National Sword policy. Some say we should pare back recycling programmes, collecting only what everyone agrees it makes sense to recycle, such as corrugated cardboard and aluminium cans.[23] That's one way to make sorting easier.

But it seems like a backward step. People in Taiwan seem to sort their own rubbish well enough, so why can't everyone else?[24] Perhaps we need systemic answers: maybe regulators can encourage new business models like those bottle-deposit schemes, making manufacturers think through the incentives and logistics for recycling their products. Those discussions are happening under the voguish phrase 'the circular economy'.[25]

Or perhaps technology will come to the rescue. One UK start-up says it can turn mixed plastics, notoriously difficult to recycle, back into the oil they came from.[26] A mall in Australia recently debuted an AI-enabled trash can which senses what you put in it and sorts accordingly; it even looks a bit like Wall-E.[27] State-of-the-art sorting facilities use lasers and magnets and air jets to separate different recyclable streams.[28]

None of this can yet compete with low-cost labourers in China – but maybe closing off that option will prove just the spur to innovation that the industry needs.

43

Dwarf Wheat

In the early 1900s, newlyweds Cathy and Cappy Jones left Connecticut to start a new life as farmers in north-west Mexico's Yaqui Valley. It was a little-known place, a few hundred kilometres south of the border with Arizona – dry, dusty and largely destitute. But, for the Joneses, it became home. They raised two daughters. When Cappy died in 1931, Cathy decided to stay on.

Just down the road, the state's ambitious governor set up an agricultural research centre – the Yaqui Valley Experiment Station. Impressive stone pillars went up at the entrance. Irrigation canals were dug. For a while, the centre raised cattle, sheep and pigs. It grew oranges, figs and grapefruit. And then it fell into disuse. By 1945 the fields were over-grown, the fences fallen, the buildings' windows shattered and roof tiles missing. The place swarmed with rats.

But then Cathy heard strange rumours: some crazy gringo had set up camp in this dilapidated place, despite the lack of electricity, or sanitation, or running water. And he hadn't brought machinery. He'd been digging with a hoe.[1]

Cathy drove over to investigate. She learned that the young

man was from Iowa, and he was working for the Rockefeller Foundation, trying to breed wheat that could resist stem rust, a disease that had ruined many a crop. Further south, where he was supposed to be based, you had to sow in spring and harvest in autumn. Up here, with a different climate, you could sow in autumn and harvest in spring. By relocating for a few months, perhaps he could find varieties that would grow in diverse conditions – and do his experiments twice as quickly.

But there was a problem. Mexico's government stipulated which regions the Foundation could work in, and this wasn't one of them. His bosses told him he could go if he wanted – but they couldn't pay for a tractor, or to make the place habitable. Officially, they'd know nothing about it. He left his wife and toddler in Mexico City, and went anyway.

Cathy took pity on the determined young Iowan. 'I couldn't have survived without Mrs Jones,' he later said. She invited him round for weekly meals, and to have a bath, and to wash his clothes. She taught him Spanish. And she drove him into the nearest town to do his shopping. Twenty-three years later, the main street of that town was renamed in his honour: Calle de Dr Norman E. Borlaug.

That same year, 1968, the Stanford biologist Paul Ehrlich published an explosive book. In *The Population Bomb*, Ehrlich noted that in poor countries such as India and Pakistan, populations were growing more quickly than food supplies. In the 1970s, he predicted, 'hundreds of millions of people are going to starve to death'.[2]

Thankfully, Ehrlich was wrong, because he didn't know what Norman Borlaug had been doing. The crazy gringo later got a Nobel Prize for the years he spent shuttling between Mexico City and the Yaqui Valley, growing thousands upon thousands of kinds of wheat, and carefully noting their traits: this kind resisted one type of stem rust, but not another; this kind produced good yields, but made bad bread; and so on. He'd cross the varieties that had some good traits, and hope that one of the cross-breeds would happen to inherit all the good traits and none of the bad.

It was painstaking work. But, eventually, it paid off. Borlaug produced new kinds of 'dwarf' wheat that resisted rust, yielded well, and – crucially – had short stems, so they didn't topple over in the wind. Through further tests, he worked out how to maximise their yield – how far apart to plant them, how deep, with how much fertiliser, and how much to irrigate.

By the 1960s, Borlaug was traversing the world to spread the news. It wasn't easy. Many couldn't conceive that another way was possible. In Pakistan, the director of a research institute sadly reported that they'd tried his wheat, but yields were poor. Borlaug saw why: ignoring his instructions, they'd planted too deep, too far apart, and without fertilising or weeding. The man replied, perplexed, 'This is the way you plant wheat in Pakistan.' For half a century, Pakistan's wheat yields had been consistent: never above 800 pounds an acre. Mexican farmers were now getting over three times that. So was Mexico's way worth a shot? No, said an eminent academic. 'These figures prove that Pakistan's wheat production will never rise!'

Borlaug could be blunt with people who didn't get it, no matter who they were. In India, he got into a yelling match with the deputy prime minister.[3] And his haranguing worked. Eventually, developing countries started to import Borlaug's seeds, and methods. And from 1960 to 2000, their wheat yields trebled. Similar work followed on corn and rice.[4] It was dubbed the 'green revolution'. Ehrlich had predicted mass starvation; in fact, the world's population more than doubled.[5]

And yet worries about overpopulation never entirely go away. Perhaps that's to be expected. It's one of the oldest questions in economics, dating back to the world's first professor of 'political economy', Thomas Robert Malthus.[6] In 1798, Malthus published *An Essay on the Principle of Population*. It made a simple argument: populations increase exponentially – two, four, eight, sixteen, thirty-two. Food production doesn't. Sooner or later, there are bound to be more people than food, with consequences that won't be pleasant.[7]

Happily for us, it turned out that Malthus had underestimated how, as people get richer, they tend to want fewer kids, so populations grow more slowly: 1968, the year that Paul Ehrlich made his dire predictions, was also the year in which population growth peaked.

Malthus also underestimated – well, Norman Borlaug. Over the years, human ingenuity has meant that food yields have kept pace.

At least, so far. But the global population is still growing.[8] By one reckoning, food yields need to keep rising at 2.4 per cent a year. And they're not: progress has slowed, and problems are mounting up – climate change; water shortages; pollution from fertilisers and pesticides. These are problems the green revolution itself has made worse. Some say it even perpetuated the poverty that keeps the population growing: fertilisers and irrigation cost money that many peasant farmers can't get.[9]

Paul Ehrlich, now in his eighties, maintains that he wasn't so much wrong as ahead of his time.[10] Perhaps if Malthus were still alive, in his two hundred and fifties, he'd say the same. As one expert put it to the author Charles Mann, 'the breeders have been pulling rabbits out of their hats for fifty years ... they're starting to run out of rabbits.'[11]

Or are they? Since genetic modification became possible,

it's mostly been about resistance to diseases, insects and herbicides – and while that does increase yields, it hasn't been the direct aim. That's starting to change.[12] And agronomists are only just beginning to explore the gene-editing tool CRISPR, which can do what Norman Borlaug did, much more quickly.[13]

As for Borlaug, he saw that his work had caused problems that weren't handled well, but asked a simple question – would you rather have imperfect ways to grow more food, or let people starve?[14] It's a question we may have to keep asking in the decades to come.

44

Solar Photovoltaics

Socrates believed that the ideal house should be warm in the winter and cool in the summer. With clarity of thought like that, it's easy to see how the great philosopher got his reputation.[1]

At the time, such a desire was easier to state than to achieve, yet many pre-modern civilisations figured out how to meet Socrates' requirements. They include the Greek city states that Socrates would have recognised; the Chinese in the seventh century BCE; and the Puebloans, the indigenous people of what is now the south-west United States. Each designed buildings to capture sunlight from the low-hanging winter sun, while maximising shade in the summer.[2]

All very elegant; but that's not the sort of solar power that will run a modern industrial economy. And millennia went by without much progress. *A Golden Thread*, a book published in 1980, celebrated clever uses of solar architecture and technology across the centuries, and urged modern economies racked by the oil shocks of the 1970s to learn from the wisdom of the ancients.

For example, parabolic mirrors – used in China 3000 years

ago – could focus the sun's rays to grill hotdogs. Solar thermal systems would use winter sun to warm air or water that could reduce heating bills. Such systems now meet about 1 per cent of global energy demand for heating.[3] It's better than nothing, but hardly a solar revolution.

A Golden Thread only briefly mentions what was, in 1980, a niche technology: the solar photovoltaic or PV cell, which uses sunlight to generate electricity. The photovoltaic effect isn't new. It was discovered in 1839 by the French scientist Edmond Becquerel. In 1883 Charles Fritts, an American engineer, built the first solid-state photovoltaic cells, and then the first rooftop solar array, in New York City.

These early cells – made from a costly element named selenium – were expensive and inefficient. It was hard to find any practical use for them at all. The physicists of the day had no real idea how they worked, either – that required the insight of a fellow named Albert Einstein in 1905. (Einstein realised that photons – packets of light – were knocking electrons out of their customary positions around the nuclei of atoms, freeing the electrons to circulate as an electric current.)

It wasn't until 1954 that researchers at Bell Labs in the US made a serendipitous breakthrough. By pure luck, they noticed that when silicon components were exposed to sunlight, they started to generate an electric current. Unlike selenium, silicon is cheap – and Bell Labs researchers reckoned it was 15 times more efficient.[4]

These new silicon PV cells were great for satellites; the American satellite Vanguard 1 was the first to use them, carrying six solar panels into orbit in 1958.[5] The sun always shines in space, and what else are you going to use to power a multimillion-dollar satellite, anyway? Yet solar PV had few heavy-duty applications on Earth itself; it was still far too costly.

Vanguard 1's solar panels produced half a watt at a cost of countless thousands of dollars. By the mid-1970s solar panels were down to $100 a watt – but that still meant $10,000 for enough panels to power a lightbulb. Yet the cost kept falling. By 2016 it was half a dollar a watt and still falling fast.[6] After millennia of slow progress, things have accelerated very suddenly.

Perhaps we should have seen this acceleration coming. In the 1930s an American aeronautical engineer named T. P. Wright carefully observed aeroplane factories at work. He published research demonstrating that the more often a particular type of aeroplane was assembled, the quicker and cheaper the next unit became. Workers would gain experience, specialised tools would be developed, and ways to save time and material would be discovered. Wright reckoned that every time accumulated production doubled, unit costs would fall by 15 per cent. He called this phenomenon 'the learning curve'.[7]

Three decades later, management consultants at Boston Consulting Group, or BCG, rediscovered Wright's rule of thumb in the case of semiconductors, and then other products too.[8] Recently, a group of economists and mathematicians at Oxford University found convincing evidence of learning-curve effects across more than 50 different products from transistors to beer – including photovoltaic cells. Sometimes the learning curve is shallow and sometimes steep, but it always seems to be there.[9]

In the case of PV cells, it's quite steep: for every doubling of output, cost falls by more than 20 per cent. And this matters because output is increasing so fast: the world produced 100 times more solar cells between 2010 and 2016 than it had before 2010.[10] Batteries – an important complement to solar PV – are also marching along a steep learning curve.

The learning curve may be a dependable fact about technology, but paradoxically, it creates a feedback loop that makes it harder to predict technological change. Popular products become cheap; cheaper products become popular.

And any new product needs somehow to get through the expensive early stages. Solar PV cells needed to be heavily subsidised at first – as they were in Germany for environmental reasons. More recently China has been willing to manufacture large quantities in order to master the technology – leading President Obama's US administration to complain that rather than being too expensive, imported solar panels had become unfairly cheap.

The ultra-low interest rates of recent years have also helped drag solar PV into the mainstream of the energy system; these low rates make it attractive to borrow and install solar panels that will then last for decades, generating electricity at almost no further cost beyond a little cleaning and maintenance.

Solar panels are particularly promising in poorer countries with underdeveloped and unreliable energy grids and plenty of sunshine during the day. When the Indian Prime Minister Narendra Modi assumed office in 2014, for example, he announced ambitious plans to build large utility-scale solar farms – but also to establish tiny grids in rural villages with little or no access to the grid.[11]

But now that solar PV has marched along the learning curve, it is competitive even in rich, well-connected areas. As early as 2012, PV projects in the sunny US states were signing deals to sell power at less than the price of electricity generated by fossil fuels.[12]

That was the sign that solar power had become a serious threat to existing fossil-fuel infrastructure – not because it's green but because it's cheap. In late 2016 in Nevada, for example, several large casino chains switched from the state utility to purchase their power from largely renewable sources. This wasn't a corporate branding exercise; it was designed to save them money, even after paying $150 million as a severance fee.[13]

Some industry observers think that solar energy is becoming so cheap that the big oil companies may go the way of Eastman Kodak, the bankrupt giant of photographic film.

Maybe this will happen sooner than we think. Or maybe not. The sun does not shine at night, and winter storage remains a big challenge. As Socrates warned us: the wisest people understand that they know nothing. But the learning curve tells us that the ultimate triumph of solar PV seems likely: it is getting cheaper as it gets more popular, and more popular as it gets cheaper. Socrates notwithstanding, that sounds like a recipe for success.

VIII

OUR ROBOT OVERLORDS

45

The Hollerith Punch-Card Machine

Amazon, Alphabet, Alibaba, Facebook, Tencent. Five of the world's ten most valuable companies, by the summer of 2019. All under 25 years old. And all got rich, in their own ways, on data.[1]

No wonder it's become common to call data the 'new oil'.[2] As recently as 2011, five of the top ten were oil companies.[3] Now only ExxonMobil clings on.

The analogy isn't perfect.[4] Data can be used many times, oil only once. But data is like oil in that the crude, unrefined stuff is not much use to anyone. You have to process it to get something valuable: diesel, to put in an engine; insights, to inform a decision. A decision such as: which advert to insert in a social media timeline; which search result to put at the top of the page.

Imagine you were asked to make just one of those decisions. Someone is watching a video on YouTube, which is run by Google, which is owned by Alphabet – what to suggest she watches next? Pique her interest, and YouTube gets to serve her another advert. Lose her attention, and she'll click away.

You have all the data you need. Look at all the other

YouTube videos she's ever watched: what is she interested in? Now look at what other users have gone on to watch after this video. Weigh up the options, calculate probabilities. If you choose wisely, and she views another ad, well done: you've earned Alphabet all of, ooh, maybe 20 cents.[5]

Clearly, relying on humans to process data would be impossibly inefficient. These business models need machines. In the data economy, power comes not from data alone, but from the interplay of data and algorithm.

In the 1880s, a young German-American inventor tried to interest his family in a machine to process data faster than humans could manage. He had designed it, but now he needed money to test it. Picture something that looks a bit like an upright piano, but instead of keys it has a slot for cards, about the size of a dollar bill, with holes punched in them; facing you are 40 dials, which may or may not tick upwards after you insert each new card.

Herman Hollerith's family didn't get it. Far from rushing to invest, they laughed at him. Hollerith evidently did not forgive: he cut them off. His children were to grow up with no idea they had relatives on their father's side.[6]

Hollerith's invention responded to a very specific problem. Every ten years, the US government conducted a census. That was nothing new. Governments through the ages have wanted to know who lives where and who owns what, to help raise taxes, and find conscripts for the army. America's Founding Fathers said the census should inform electoral boundaries, so that every area had equal representation in Congress.[7]

But if you're going to send a small army of enumerators around the country, it must be tempting to ask about an ever-wider range of things. What jobs do people do? Any illnesses or disabilities? What languages do they speak? Knowledge is power, as nineteenth-century bureaucrats knew just as well as twenty-first-century platform companies. Yet with the 1880 census, the bureaucrats had swallowed more data than they could digest. The census had been expanding to include libraries, nursing homes, crime statistics and many other topics. In 1870, the census had five different kinds of questionnaire. In 1880, it had 215.[8] It soon became clear that adding up the answers would take years: they'd barely have finished this census when it would be time to start the next one. A lucrative government contract surely awaited anyone who could speed the process up.

Young Herman had worked on the 1880 census, so he understood the problem. He'd also worked for the patent office, where he'd seen that there was money in inventing – especially for the fast-growing railways.[9] Herman had decided to seek his fortune by inventing a new kind of brake for trains. As it happened, a train journey helped him to solve the census problem instead.

Tickets were often stolen, so railway companies found an ingenious way to link them to the person who'd bought them: a 'punch photograph'. Conductors used a hole-punch to select from a range of physical descriptors: as Hollerith recalled, 'light hair, dark eyes, large nose etc.'[10] If a dark-haired, small-nosed scoundrel stole your ticket, he wouldn't get far.

But after observing this system, Hollerith realised that people's answers to census questions could also be represented as holes in cards. That could solve the problem, because punched cards had been used to control machines since the early 1800s: the Jacquard loom wove patterned fabric based on them. All Hollerith needed to do was make a 'tabulating machine' to add up the census punch cards he envisaged. In that piano-like contraption, a set of spring-loaded pins descended on the card; where they found a hole they completed an electrical circuit, which moved the appropriate dial up by one.

Happily for Hollerith, the bureaucrats were more impressed than his family. They rented his machines to count the 1890 census, to which they'd added yet more questionnaires.[11] The paperwork weighed about 200 tons.[12] Compared with the old system, Hollerith's machines proved years quicker and millions of dollars cheaper.[13]

More importantly, they made it easier to interrogate the data. Suppose you wanted to find people aged 40 to 45, married, and working as a carpenter.[14] No need to sift through 200 tons of paperwork – just set up the machine, and run the cards through it. 'By simple use of the well-known electrical relay,' Hollerith explained, 'we can secure any possible combination.'[15]

Governments soon saw uses far beyond the census: 'Across the world,' says the historian Adam Tooze, 'bureaucrats were inspired to dream of omniscience.'[16] America's first social-security benefits were disbursed through punched cards in the 1930s.[17] The following decade, punched cards notoriously helped organise the Holocaust.[18]

Businesses, too, were quick to see the potential. Insurers used punched cards for actuarial calculations, utilities for billing, railways for shipping, manufacturers to keep track of sales and costs.[19] Hollerith's Tabulating Machine Company did a roaring trade. You may have heard of the firm that, through mergers, it eventually became: IBM remained a market leader as punched cards gave way to magnetic storage, and tabulating machines to programmable computers. It was still on the list of the world's ten biggest companies a few years ago.[20]

But if the power of data was apparent to Hollerith's customers, why did the data economy take another century to arrive? Because there's something new about the kind of data that's now being compared to oil: the likes of Google and Amazon don't need an army of enumerators to collect it. We trail it behind us every time we use our phones, or ask Alexa to turn the light on.

This kind of data is not as neatly structured as the predefined answers to census questions precision-punched into Hollerith's cards. That makes it harder to make sense of. But there's unimaginably more of it. And as algorithms improve, and more of our lives are lived online, the bureaucratic dream is fast becoming corporate reality.

46

The Gyroscope

On 3 October 1744, a storm was brewing in the English Channel. With sails set for home after chasing a French fleet off the coast of Portugal, a squadron of English warships led by Admiral Sir John Balchen was heading straight for trouble. '[W]e met with a hard Gale of Wind which tore all our Sails and Rigging that we were obliged to submit to the Mercy of the Waves,' wrote one sailor whose ship made it back to harbour – but only just. 'On the 4th we had Ten Feet in our Hold, which made our Condition very bad, and the Dread of Death appeared in every Face, for we momentarily expected to be swallowed up.'[1]

One ship *was* swallowed up – the flagship: HMS *Victory*, commanded by Admiral Balchen himself. It sank a hundred metres to the seabed 50 miles south of Plymouth, taking with it eleven hundred men and, so rumour had it, quite a lot of Portuguese gold bullion.[2] There the wreckage lay until 2008, when treasure hunters located it. They were hoping to find the gold – but there was something else on board that ship that has proved much more economically significant: the first known attempt to apply an idea that's now used to guide everything from submarines to satellites, from rovers on Mars to the phone in your pocket.

The man who had that idea was called John Serson, and a year earlier he'd been invited on board a royal yacht near London to explain it to two high-ranking naval officers and an eminent mathematician. Serson was a sea captain. He was barely literate. But he was an 'ingenious mechanick', as *The Gentleman's Magazine* later put it. Serson's idea was inspired by a child's toy – a spinning top. The problem he wanted to solve was this: sailors worked out a ship's position by using a quadrant to take an angle from the sun to the horizon – but you couldn't always see the horizon, for haze or mist.

Serson wondered if he could create an artificial horizon – something that would stay level, even as a ship lurched and swayed around it. As *The Gentleman's Magazine* recounts, he:

> got a kind of top made, whose upper surface perpendicular to the axe was a circular plane of polish'd metal; and found, as he had expected, that when this top was briskly set in motion, its plane surface would soon become horizontal ... if the whirling plane were disturbed from its horizontal position, it would soon recover it again.[3]

The officers and the mathematician were impressed: 'in their opinion, Mr Serson's contrivance was highly deserving their encouragement, as likely to prove very useful in foggy weather'.[4] The navy asked Serson to make further observations, aboard the HMS *Victory*, 'and so perish'd poor Mr Serson'.[5] But his idea lived on. Others made versions of his device.[6] One was sold to the French Academy of Sciences, much to the disdain of *The Gentleman's Magazine*: ''Tis not improbable, therefore, that the French may make some insignificant alterations in it, and in time, as usual, venture to call it their own.'[7]

As it turned out, insignificant alterations weren't enough: Serson's 'whirling speculum' proved of sadly limited practical use.[8] But it was France, a century later, that gave us a more successful take on the same principle. This was a spinning disc mounted in gimbals, which are a set of pivoted supports that allow the disc to maintain its orientation regardless of how the base might be tilting around.

The physicist Léon Foucault called his device a gyroscope, from the Greek words for 'turn' and 'observe', because he used it to study the Earth's rotation. Then electric motors came along, meaning that the disc could spin indefinitely. And practical applications came thick and fast. Ships got workable artificial horizons. So did aeroplanes. In the early 1900s, Hermann Anschütz-Kaempfe and Elmer Sperry figured out how to align the spin to the Earth's north–south axis, giving us the gyrocompass.

Combine these instruments with others – accelerometers, magnetometers – and you get a good idea of which way up you are and in which direction you're heading. Feed the outputs into systems that can course-correct, and you have an aeroplane's autopilot, a ship's gyro-stabiliser, and inertial navigation systems on spacecraft or missiles.[9] Add in GPS, and you know where you are.

There's a limit to how small you can make spinning discs in gimbals, but other technologies have miniaturised the gyroscope.[10] Vibrating micro-electro-mechanical gyroscopes measure just a few cubic millimetres.[11] Researchers are making a laser-based gyroscope thinner than a human hair.[12] As these and other sensors have got smaller and cheaper – and computers faster, and batteries lighter – they've found uses from smartphones to robots, gaming consoles and virtual-reality headsets.

And another technology around which there's a particular buzz: the drone.

The first pilotless flight is often traced to 1849 – just three years before Léon Foucault's gyroscope. Austria tried to attack Venice by fixing bombs to balloons and waiting for the wind to blow in the right direction.[13] This was not a triumph: some bombs landed in Austrian territory.[14] But military uses continued to drive drone technology, until very recently: search for 'drones' in a news archive, and you'll find that the top stories were about war until about four or five years ago. Then suddenly they start being about 'What do airspace regulations mean for hobbyists?', and 'How long before drones are delivering our groceries?'

That's a big question. Drones are now commonplace from surveying to movie-making; they get urgent medical supplies to hard-to-reach places. But it's the routine, everyday uses that promise to be truly transformative: flying our online shopping to us, or even flying *us* – the Chinese company Ehang is pioneering drones that can carry human passengers.[15]

In rural China, delivery drones are starting to look like a leapfrog technology: one that catches on most quickly where there isn't a competing established infrastructure – in this case, of big-box retail stores and roads for van deliveries. Zhangwei, for example, is a village in Jiangsu province where few people own cars, and only half have fridges, but everyone has a phone – and they use their phones to place orders at online retailer JD.com, for everything from disposable nappies to fresh crabs. About four times a day, warehouse workers dispatch the village's orders on a drone that carries up to 14 kilograms at 45 miles an hour. Everyone's happy – except for the woman who runs the village shop.[16]

If drones are to take off more widely, we'll need solutions to the 'last mile' problem. In Zhangwei, JD.com employs a human to distribute the crabs and the nappies to the villagers who ordered them – but in countries where labour is pricier, the last mile is where delivery costs are concentrated; automate it, and some believe bricks-and-mortar stores could cease to exist altogether.[17] But nobody is sure precisely how that might work. Do we want our online purchases parachuted into our back gardens, or plunked on the roofs of our apartment buildings? How about smart windows that can open to let drones in when we're not at home?[18]

Then there's another problem – the one that did for poor John Serson: the weather. If we're going to rely on airborne deliveries, they'll have to work in all conditions.[19] Will drones ever navigate storms that could sink a battleship? Perhaps then the promise of the gyroscope will have truly been fulfilled.

47

Spreadsheets

In 1978, a Harvard Business School student named Dan Bricklin was sitting in a classroom, watching his accounting lecturer filling in rows and columns on the blackboard. The lecturer made a change, and then had to work down and across the grid on the board, erasing and rewriting other numbers to make everything add up. That looked like boring, repetitive hard work to Dan Bricklin – and laziness can be the mother of invention.[1]

The lecturer wasn't the only person writing down numbers in rows and columns, then taking ages to rub them out and recalculate: accounting clerks all over the world did that every day in the pages of their ledgers. A two-page spread across the open fold of the ledger was called a 'spreadsheet'. The output of several paper spreadsheets would then be the input for some larger, master spreadsheet. Making an alteration might require hours of work with a pencil, eraser, and desk calculator.

Like many business-school students, Bricklin had had a real job before going to Harvard – he'd worked as a programmer at Wang and DEC, two big players in 1970s computing. And he thought: why on earth would anyone do this on a blackboard, or on a paper ledger, when you could do it on a computer?

So he wrote a program for the new Apple II personal computer: an electronic spreadsheet. A friend of his, Bob Frankston, helped him sharpen up the software, and on 17 October 1979 their brainchild, VisiCalc, went on sale. Almost overnight, it was a sensation.

Other financial and accounting software had long existed, but VisiCalc was the first program with the modern spreadsheet interface – and it is widely thought to be the first 'killer app', a software program so essential that you'd buy a computer just to be able to use it. Steve Jobs of Apple later said that it had been VisiCalc that 'propelled the Apple II to the success it achieved'.[2]

Five years after VisiCalc's launch, the journalist Steven Levy – an unofficial historian of modern computing – was able to write, 'There are corporate executives, wholesalers, retailers, and small business owners who talk about their business lives in two time periods: before and after the electronic spreadsheet.'[3]

Levy also noted the appearance of a new, powerful rival to VisiCalc, Lotus 1-2-3. And by 1988, the *New York Times* reported that 'Lotus has dominated the spreadsheet market for five years' after it 'toppled the first spreadsheet, VisiCalc, whose dominant share of the personal computer market seemed invincible'. How are the mighty humbled! The *New York Times* also described several upstart challengers, including a program called Microsoft Excel.[4]

But the real lesson of the spreadsheet is not about how monopolies rise and fall, but about technological unemployment. The cliché these days is that the robots are coming for our jobs. But the story is never as simple as that, and the best illustration I can think of is the digital spreadsheet.

What does a robot accountant look like, after all? It certainly isn't Arnold Schwarzenegger's Terminator, equipped with a pocket calculator instead of a machine gun. Admittedly, if I were a human accountant and came into work one morning to find Arnie sitting at my desk, I'd back out stealthily, resolving to collect my personal effects later.

If the concept of a robot accountant means anything, surely it means VisiCalc or Excel. These programs put hundreds of thousands of accounting clerks out of work. Accounting clerks were the men and women who spent their days tapping away at pocket calculators while erasing and recalculating numbers on paper ledger sheets. Of *course* VisiCalc was revolutionary in that world. Of *course* it was more efficient than a human. According to the Planet Money podcast, in the US alone, 400,000 fewer accounting clerks are employed today than in 1980, the first full year that VisiCalc went on sale.

But Planet Money also found that there were 600,000 *more* jobs for regular accountants. After all, crunching the numbers had become cheaper, more versatile, and more powerful, so demand went up. The point is not really whether 600,000 is more than 400,000: sometimes automation creates jobs and sometimes it destroys them. The point is that automation reshapes the workplace in much subtler ways than 'a robot took my job'.

In the age of the spreadsheet, the repetitive, routine parts of accountancy disappeared. What remained and indeed flourished required more judgement, more human skills. The spreadsheet created whole new industries. There are countless jobs in high finance that for the purposes of trading, or insuring, or whatever, depend on exploring different numerical scenarios – tweaking the numbers and watching the columns recalculate themselves. These jobs barely existed before the electronic spreadsheet.

In *Fifty Things That Made the Modern Economy*, we encountered the Jennifer Unit, an earpiece that directs warehouse pickers to collect products by breaking down instructions into the most mindless, idiot-proof steps. The Jennifer Unit strips a menial task of its last faintly interesting element. The spreadsheet operates in reverse: it strips an intellectually demanding job of the most boring bits.

Viewed together, two technologies show that technology doesn't usually take jobs wholesale – it chisels away easily automated chunks, leaving humans to adapt to the rest. That can make the human job more interesting, or more soul-destroying – it all depends.

In accountancy, it made the human jobs more creative. Who doesn't want a creative accountant? Far from being traumatised by automation, accountants may now take the spreadsheet for granted. The histories of accountancy that I've read don't bother to mention VisiCalc or Excel. Perhaps it seems beneath their dignity.

What the spreadsheet did to accounting and finance is a harbinger of what is coming to other white-collar jobs. Journalists no longer churn out routine stories about corporate earnings reports; algorithms do that more quickly and cheaply. Teachers give help to pupils after an online tutorial has quizzed the child and figured out where she is getting stuck. A doctor can sometimes be replaced by a combination of a nurse and a diagnostic app. Law firms use 'document assembly systems' that quiz clients and then draft legal contracts.[5] Whether members of these professions will look back as kindly as accountants on their encounter with Arnie the Terminator remains to be seen.

But they should learn one final cautionary tale the spreadsheet has to offer: sometimes we think we have delegated some routine job to an infallible computer, whereas we've simply acquired a lever with which to magnify human error to a dramatic scale.

Consider the time when unsuccessful applicants for a senior police job were told they'd been offered the job: that's what happens when you sort one column without sorting the adjacent one.[6]

Or the time two noted economists, Carmen Reinhart and the former IMF Chief Economist Ken Rogoff, were mightily embarrassed when a graduate student spotted a spreadsheet error in an influential economics paper. Reinhart and Rogoff accidentally omitted several countries because they forgot to drag down the formula selection box by five more cells.[7]

Oh, and there's the time that the investment bank J. P. Morgan lost $6 billion, in part because a risk indicator in a spreadsheet was being divided not by an average of two numbers, but by their sum – making the risks look half as big as they should have done.[8]

If we ask computers to do the wrong thing, they'll do it with the same breath-taking speed and efficiency that inspired Dan Bricklin to create VisiCalc. That is a lesson we seem doomed to keep learning far beyond the borders of accountancy.

48

The Chatbot

Robert Epstein was looking for love. The year being 2006, he was looking online. He began a promising email exchange with a pretty brunette, but before long he realised he'd been deceived. In halting English, Ivana admitted that she didn't live nearby in California, but in Russia. Epstein was disappointed – he wanted more than a penfriend, let's be frank – but she was warm and friendly. Soon she confessed that she was developing a crush on him.[1]

'I have very special feelings about you . . . It – in the same way as the beautiful flower blossoming in mine soul . . . I only cannot explain . . . I shall wait your answer, holding my fingers have crossed . . . '

The correspondence blossomed. It took a long while for Epstein to notice that Ivana never really responded directly to his questions. She'd write about taking a walk in the park, having conversations with her mother, and repeat sweet nothings about how much she liked him. Suspicious, he eventually sent Ivana a line of pure bang-on-the-keyboard gibberish. She responded with another email about her mother. At last, Robert Epstein realised the truth: Ivana was a chatbot.

What makes the story surprising isn't that a Russian chatbot managed to trick a lonely middle-aged Californian man. It's that the man who was tricked was one of the founders of the Loebner Prize, an annual test of artificial conversation in which computers try to fool humans into thinking that they, too, are human. One of the world's top chatbot experts had spent two months trying to seduce a computer program.

Each year, the Loebner Prize challenges chatbots to pass the Turing test, which was proposed in 1950 by the British mathematician, codebreaker and computer pioneer Alan Turing. In Turing's 'imitation game', a judge would communicate through a teleprompter with a human and a computer.[2] The human's job was to prove that she was, indeed, human. The computer's job was to imitate human conversation convincingly enough to persuade the judge.[3]

There is a long history of computing pioneers being over-optimistic about the rise of the machines. Herb Simon, later a Nobel laureate in economics, predicted in 1957 that a computer would beat the world chess champion within ten years; it took forty – a topic to which we will return in the final chapter. In 1970 Marvin Minsky predicted that computers would have human-like general intelligence 'within three to eight years'. That now seems absurd.

Alan Turing's prediction fared a bit better. He thought that within 50 years, computers would be able to fool 30 per cent of human judges after five minutes of conversation. Not far off. It took 64 years – although we're still arguing over whether 'Eugene Goostman', the program that in 2014 was trumpeted as passing the Turing test, really counts.[4] Like Ivana, Goostman managed expectations by claiming not to be a native English speaker. (He said he was a 13-year-old kid from Odessa in Ukraine.)

One of the first and most famous early chatbots, ELIZA,

would not have passed the Turing test – but did, with just a few lines of code, successfully imitate a human non-directional therapist. ELIZA was named after the fictional character Eliza Doolittle, of *Pygmalion* and *My Fair Lady*. She – it? – was programmed by Joseph Weizenbaum in the mid-1960s. If you typed, 'My husband made me come here' ELIZA might simply reply, 'Your husband made you come here.' If you mentioned feeling angry, ELIZA might ask, 'Do you think coming here will help you not to feel angry?' Or she might simply say, 'Please go on.'

People didn't care that ELIZA wasn't human: at least someone would listen to them without judging them or trying to sleep with them. Joseph Weizenbaum's secretary asked him to leave the room so that she could talk to ELIZA in private.[5]

Psychotherapists were fascinated. 'Several hundred patients an hour could be handled by a computer system,' mused a contemporary article in the *Journal of Nervous and Mental Disease*. Supervising an army of bots, the human therapist would be far more efficient.[6] And indeed, cognitive behavioural therapy is now administered by chatbots, such as 'Woebot', designed by a clinical psychologist, Alison Darcy. There is no pretence that they are human.[7]

Joseph Weizenbaum himself was horrified at the idea that people would settle for so poor a substitute for human interaction. But like Mary Shelley's Frankenstein, he had created something beyond his control.

Chatbots are now ubiquitous. They handle complaints and enquiries. Babylon Health is a chatbot that quizzes people about their medical symptoms and decides whether they should be referred to a doctor. Amelia talks directly to the customers of some banks, but is used by Allstate Insurance to provide information to the call-centre workers who themselves are talking to customers. And Alexa and Siri interpret our voices and speak back, with the simple goal of sparing us stabbing clumsily at tiny screens.[8]

Brian Christian, the author of *The Most Human Human*, a book about the Turing test, points out that most modern chatbots don't even try to pass it. There are exceptions: Ivana-esque chatbots were used by Ashley Madison, a website designed to facilitate extramarital affairs, to hide the fact that few human women used the site.[9] It seems we're less likely to notice a chatbot isn't human when they plug directly into our libido.

Another tactic is to rile us up. One effective chatbot, MGonz, tricks people by starting an exchange of insults.[10] Politics – most notoriously the 2016 US election campaign – is well-seasoned with social media chatbots pretending to be outraged citizens, tweeting lies and insulting memes.[11]

But generally chatbots are happy to present as chatbots. Seeming human is hard. Commercial bots have largely ignored the challenge, and specialised in doing small tasks well – solving straightforward problems, and handing off the complex cases to a person. Adam Smith explained in the late 1700s that productivity is built on a process of dividing up labour into small specialised tasks.[12] Modern chatbots work on the same principle.

The logic leads economists to believe that automation reshapes jobs rather than destroying them. As we saw with the spreadsheet, jobs are sliced into tasks. Computers take over the routine tasks. Humans supply the creativity and the adaptability.[13] That is what we observe from the cash machine to the self-checkout kiosk. Chatbots give us another example.

But we must be wary of the risk that as consumers, pro-ducers – and perhaps even as ordinary citizens – we contort ourselves to fit the computers. We use the self-checkout even though a chat with a shop assistant might lift our mood. We post status updates – or just click an emoji – that are filtered by social media algorithms; as with ELIZA, we're settling for the feeling that someone is listening.[14]

Brian Christian argues that we humans should view this as a challenge to raise our game. Let the computers take over the call centres. Isn't that better than forcing a robot made of flesh-and-blood to stick to a script, frustrating everyone involved? We might hope that rather than fooling humans, better chatbots will save time – freeing us up to talk more meaningfully to each other.

49

The CubeSat

There's a popular and well-loved story about the dimensions of the Space Shuttle. Apparently the booster rockets
had to fit through railway tunnels whose dimensions were
influenced by the size of a horse and cart: in short, the space
shuttle boosters were the width of two horses' backsides.

That tale is probably a little tenuous, but a similar – and
quite true – story can be told about the new poster child of
the space industry. Its dimensions were determined by the
size of a Beanie Baby.[1]

Beanie Babies were all the rage in 1999, at a time when
Stanford University Professor Bob Twiggs was teaching his
graduate students to design satellites. Back then, satellites
were big. For example, the Artemis telecommunications
satellite, launched in 2001, weighed more than 3 tonnes, was
8 metres tall, and each of its two solar panels was as long as
a bus.[2] With that much space and weight to play with, the
temptation was to pack more and more gear into the satellite,
making it more and more expensive – not to mention an
encouragement to lazy thinking.

'If you've got lots of room to put everything in, you end

up not being too careful with it,' says Twiggs.[3] So he and his colleague decided that the students needed a constraint. Twiggs went to the local store, where he spotted . . . a Beanie Baby neatly packed in its box. He went back to class, placed the Beanie Baby box on the desk, and told his students: your satellite needs to be able to fit in this box.[4]

As modern smartphones have revolutionised the quality and power of small off-the-shelf components, this educational challenge has evolved into a practical standard for tiny satellites – the CubeSat. 'CubeSat' is a slight misnomer: the unit is 10cm by 10cm by 11.35cm, and many CubeSats are several units big – but still about the size of a shoebox, kilograms rather than tonnes.

One planned CubeSat, the Lunar Flashlight, aims to orbit the moon, reflecting sunlight into deep craters and analysing the light that bounces back. Another project, the Near-Earth Asteroid Scout, is designed to demonstrate a solar sail while exploring nearby asteroids.[5]

But for now, most CubeSats are designed to take photographs and other images of our planet from above. The basic ingredients: a smartphone processor, solar panels, a camera and some batteries.[6]

CubeSats are cheap to make and cheap to launch. Traditionally, the entire process of building and launching a major satellite might cost $500 million. You could get a CubeSat into low earth orbit for closer to $100,000.[7]

Big rockets such as the European Space Agency's Ariane 5, or Russia's Soyuz 2, are about 50 metres tall. But CubeSats and other tiny satellites can ride on much smaller private-sector rockets – say, the 18-metre Electron rocket from the New Zealand Launchpad of Rocket Labs.

CubeSats can also piggyback on a large satellite launch. Early in 2017, India's official space agency, ISRO, launched

104 satellites in a single launch – a world record. Three of the satellites were large, but the rest were tiny. Eighty-eight of them were CubeSats owned by a new Silicon Valley company, Planet.[8]

Planet was founded in 2010 and has the world's largest private fleet of satellites – around 140 of them as of the summer of 2019, taking 800,000 photographs a day – covering anywhere on the globe, once every 24 hours. They can't match the sophisticated imaging of a large satellite, but they make up for that by being able to provide better coverage – more photographs of more places within any given time frame. And Planet's 140 satellites may be the vanguard of something much bigger: both SpaceX and Amazon have announced plans to launch thousands of satellites in low earth orbit.[9]

CubeSats have three lessons to teach us about the modern economy. First, the importance of cheap, standardised modular components. While we reserve our attention and our plaudits for unique and complex projects, being cheap changes everything.

Second, CubeSat pioneers have embraced the fail-fast model of Silicon Valley. NASA, as a public agency, has a very low tolerance for risk. But an expendable CubeSat allows a different approach: if you're launching dozens at a time, you can lose one or two here and there. While NASA was focusing on ensuring that expensive kit worked perfectly, the Silicon Valley model of the CubeSats says: don't worry. Failing with disposable satellites is cheaper than succeeding with big ones. If it doesn't work, try again.

But lesson three, don't dismiss the public sector too casually. It's easy to define private space exploration in contrast with NASA and other national space agencies – in fact, I just did. But NASA has quietly supported CubeSats – for example, by funding small CubeSat-launching rockets, and by giving CubeSats free rides to the International Space Station, where they can be launched through a special CubeSat airlock.[10]

CubeSats may soon be teaching us something entirely new about the way the economy works. The great economist Alfred Marshall, who died in 1924, described economics as being the study of humanity 'in the ordinary business of life'. CubeSats allow us to observe the ordinary business of life as it unfolds, all around the world, day by day, and in some detail.

Economic forecasters haven't been slow to notice this possibility. Lots of people would love to know whether the price of oil is likely to go up or down, whether there's a glut in the market for wheat, or a shortage of high-quality Ethiopian coffee: commodity traders; crop insurers; supermarkets; oil companies; even Starbucks. It doesn't take much imagination to see how daily images of crops would give you an edge – and with the right analysis, and the right photographs, you may also be able to spot trucks on the road, count oil-storage tanks, or even how much electricity a power plant is generating by looking at the plumes of smoke.[11]

But beyond these narrow trading forecasts, satellites promise to illuminate hidden connections in the way the world economy works. We can measure pollution, congestion, deforestation, even attempts at ethnic cleansing.[12] Algorithms are starting to extract subtle information at scale: how many of those houses in a Kenyan village have metal roofs? Which roads in Cameroon are in good condition – and has foreign aid money made any difference?[13]

There's so much going on under the surface of a big economy, and so much that doesn't show up in regular statistical releases for months – sometimes years. Now we can see it day by day.

As the old story about the Space Shuttle and the horse's backside reminds us, some things in our economy change slowly. But a lot of the modern economy moves very quickly indeed. No wonder some people are keen to take snapshots.

50

The Slot Machine

Mollie's first job, as a young teenager, was dispensing change for slot machines on a military base. By the time she hit middle age, Mollie was no longer earning her wages from the slots – she was feeding her entire paycheque into them in two-day binges.[1]

'I even cashed in my life insurance for more money to play,' she told Natasha Dow Schüll in a hotel room high above the Las Vegas strip. Schüll is an anthropologist who has been studying the world of slot machines for two decades.

Perhaps it's appropriate that the conversation took place between two women. Sociologists have often described gambling as a proof of *manliness* – from a James Bond in tuxedo demonstrating his nerves of steel at high-stakes roulette and his skill at poker, to the cockfighting gamblers of Bali analysed by the anthropologist Clifford Geertz in the 1970s. Slot machines don't seem to fit at all. They require neither skill nor steely nerves. Geertz argued that they were a diversion for 'women, children, adolescents . . . the extremely poor, the socially despised, and the personally idiosyncratic'.[2]

But slot machines are no toy. They are fantastically

profitable, and they have grown like an invasive species. I encountered them en masse in 2005, when I travelled to Las Vegas to write about game theory at poker's World Series. Dozens of journalists jostled to interview star players. The slot machines seemed a depressing but colourful decorative backdrop, cocooning obese and elderly players who would ride them like motorised wheelchairs. It was only later that I realised that it was the World Series of Poker that was the decorative backdrop. As far as the casinos were concerned, the slot machines had become the main event.[3]

Not just the casinos either – the UK's gambling industry, once dominated by betting on horse racing, has become dependent on a kind of slot machine called a Fixed Odds Betting Terminal. When the government announced in 2018 that maximum bet sizes would be cut, one bookmaker responded by saying it would have to close nearly 1000 shops.[4]

Mollie spends so much on the slot machines that a Vegas hotel has invited her to stay there free of charge. Is Mollie hoping for a big win?, Natasha Dow Schüll asks. No. She knows there's no chance of that.

'The thing people never understand is that *I'm not playing to win.*'[5]

A gambler who doesn't care about winning? That doesn't seem right. But we've long struggled to appreciate what slot machines really are, and the lesson they have to teach us about the modern economy.

Slot machines are generally reckoned to have begun in the US around 1890. The Ideal Toy Company of Chicago made one with five spinning drums, each with ten playing cards. Insert a coin, and if five cards line up into a decent poker hand you could collect a prize from an attendant. A Brooklyn firm, Sittman and Pitt, made a version in 1893 that was popular across the US.

Then Charles Fey – an immigrant to San Francisco from Bavaria – hit upon the idea of simplifying the device. With just three reels, the mechanism became straightforward enough that the machine would pay out without the need for a human attendant. The machine was a hit in San Francisco, until Fey's workshop was destroyed in a fire in the aftermath of the 1906 earthquake.[6]

Modern slot machines are simply computers in shells, with their chunky levers designed to evoke the old mechanical machines. It is this digital shift that has made the slot machines so profitable. No need to worry about making change – the job that teenage Mollie used to have – because players carry digital cards on lanyards that connect them umbilically to the machines. The players need never move; they enter what Mollie calls 'the zone', a trancelike state of absorption where the rest of the world dissolves. Winning simply means more credit, and more credit means more 'T.O.D.' – the acronym for the term of art, 'time on device'.[7]

That's what Mollie was talking about when she said she wasn't playing to win. Modern slot machines are not like lotteries or roulette, with players living in hope of the jackpot. Instead, the slots gulp down low stakes – perhaps a hundred one-cent bets, spread across a dizzying grid of possible winning combinations – and they constantly spit back small wins, too. If, indeed, they should be described as wins. If you've placed a hundred one-cent bets, and win back 20 cents, is that really a win? With flashing lights and celebratory jingles, the machine will tell you it is.

In one machine studied by researchers, a hundred spins would be expected to produce 14 real wins – in which the machine paid back more than the punter put in – and 18 false wins, in which the player received something with great fanfare, but less than he or she had wagered.[8] The same research team went on to demonstrate in laboratory experiments that a machine with that 18 per cent rate of false wins was more addictive than machines with far more or far fewer false wins.[9]

The slot-machine designers aren't doing this stuff by accident: the slot-machine industry is ferociously competitive. A $10,000 machine can pay for itself in a month, if it attracts the players. If not, it will be replaced – with a machine sporting a popcorn kettle in which lottery balls bubble over, or a machine that wafts the smell of chocolate in the face of the player – or a machine that in the voice of Donald Trump, announces, 'You're fired!' Anything to delight and to surprise. They are always on the lookout to build a better mousetrap, and we're the mice.[10]

B. F. Skinner, one of the most famous psychologists of the twentieth century, would not have been surprised. At Harvard University, Skinner used to investigate behaviour by giving rats who pressed a lever the reward of a food pellet. Once, trying to eke out a supply of food pellets, he gave the reward intermittently. Often the pellet would come, often not. There was no way for the rat to know. Surprisingly, the unpredictable reward was more motivating than a generous and reliable pay-off.[11]

Slot addicts such as Mollie are similarly hooked, absorbed in 'the zone'. The anthropologist Natasha Dow Schüll once watched footage, captured on a casino security camera, of someone having a heart attack at a slot machine:

He ... collapses suddenly onto the person next to him, who doesn't react at all ... two passers by stretch him out, one of them an off-duty ER nurse. Few gamblers in the immediate vicinity move from their seats. ... In less than one minute, a security officer appears on the scene bearing a defibrillator; he applies the pads, clears, and shocks the man twice ... Despite the unconscious man lying quite literally at their feet, touching the bottoms of their chairs, the other gamblers keep playing.

Research suggests that slot machines can create addicts far more quickly than other forms of gambling such as lotteries, casino games or sports betting.[12]

But just as unnerving is the sense that in the past few years, the psychology of the slot machine has escaped the casino and migrated to our pockets. Recovering addicts avoid going to places where they might see slot machines – but there is nowhere that we can escape our phones, and plenty of good reasons to start looking at them. We all see people 'in the zone', oblivious to their companions or the traffic because the phone is all that matters.

It's that intermittent reinforcement again: is there any more email? Any 'likes' on Facebook? Many computer games are more brazen in their use of intermittent reinforcement, offering 'loot boxes' with those familiar sparkles and unpredictable rewards. It looks a lot like a gambling – often underage gambling, at that.[13]

A 2003 book, *Something for Nothing*, begins with the shocking image of slot-machine gamblers urinating into cups because they don't want to break their streak on the game.[14] But these days most of us know what it's like to urinate while looking at our phones. It's not just me, is it?

We may not be looking to maximise 'time on device', but the big tech companies who make their money from advertising certainly are. The more we look at our screens, the more advertisements they can show us. Most of us will never find ourselves in Mollie's position, enslaved by slot machines. It is a shame we can't say the same thing about the shiny devices in our pockets.

51

Chess Algorithms

On 25 June 2012 Garry Kasparov – regarded by many as the greatest player in the history of chess – sat down to play a game against a computer. He wasn't sitting down for long. Despite letting the computer have the advantage of the white pieces, Kasparov was soon chasing its king with a knight, both bishops and the queen. Checkmate took 16 moves, and a mere 40 seconds, leaving Kasparov apologising for winning so quickly.[1]

Yet Kasparov was magnanimous in his praise of the computer program. It was called 'TuroChamp'. It had been written in 1948 by the mathematician Alan Turing – he of the Turing test discussed in Chapter 48. Turing had specified some simple rules: positions with more pieces, more mobile pieces, and better defended pieces were rated as preferable in Turing's system. The program simply looked at every possible move and response – typically a few hundred options – and made the move that produced the position with the highest value, assuming the opponent then delivered the most damaging reply.

On a modern laptop that calculation takes a fraction of a

fraction of a second. But Alan Turing had no computer, and took half an hour per move to make the calculations using a pencil and paper.[2] Kasparov was full of praise for a computer chess algorithm that ran without a computer.[3]*

An algorithm is a step-by-step procedure, a series of well-defined instructions that one follows to produce a result – a recipe written by an infinitely pedantic chef. These days we think of algorithms as something rather mysterious that computers do. But as TuroChamp demonstrates, algorithms are actually a way to demystify the processes needed to produce a result, so that others can do it too. Turing could have played a far better game of chess by relying on his intuition, and with far less effort. But he could never have explained quite how.

The word 'algorithm' derives from the name of a brilliant Persian mathematician active about 1200 years ago. His name was Muhammad ibn Musa al-Khwarizmi, and European scholars later called him Algorithmi.

Algorithms themselves pre-date al-Khwarizmi; they were used in Babylon nearly 4000 years ago to calculate the solutions to algebraic problems. The computer scientist Donald Knuth republished some of these ancient algorithms in 1972 to remind his colleagues that programs are far older than computers.[4] One of the algorithms that Knuth highlighted shows how to calculate the length and width of a rectangular cistern – presumably the kind of thing the ancient Babylonians might want to do – given certain facts about its depth, volume and the relationship of its width to its height.

---

* Turing wasn't the first person to write a computer algorithm for a computer that didn't yet exist. Ada Lovelace wrote one in 1843, designed to run on Charles Babbage's Analytical Engine, a mechanical computer that was never built. Lady Lovelace's program anticipates ideas such as loops and variables. It even contains a bug.

The Babylonian algorithm was basically a recipe for solving a high-school algebra problem.*

It wasn't just the Babylonians. Various algorithms were developed around the world: some of the ones we know about come from China in the third century, India in the seventh century, and of course ancient Greece. More than two thousand years ago Euclid published an algorithm for producing the greatest common divisor of two numbers. Euclid's algorithm is a simple piece of arithmetic that you perform over and over again until you converge on the answer.[5]

All of these algorithms, however, dealt with problems that were essentially mathematical, such as finding prime numbers or solving linear equations. It was in the 1850s that George Boole, professor of mathematics at Queen's College, Cork, in Ireland, published *The Laws of Thought*. Boole's book turned logical propositions into mathematical operations: TRUE, FALSE, AND, OR, NOT, which suggested the prospect of turning thought itself into a step-by-step algorithmic process. But Boole's ideas languished for 80 years; it wasn't clear what practical value they might have.

Then in the 1930s, the American mathematician Claude Shannon showed that Boole's 'laws of thought' could be obeyed by electrical circuits – TRUE or FALSE became ON or OFF. AND, NOT and OR were operations performed by simple electronic components. The digital age was arriving, and that meant that algorithms could start to realise their full potential.[6]

Chess has been a laboratory for algorithmic intelligence since the earliest days of computer science: it seemed sufficiently well defined to feel like a plausible challenge, yet too complex to solve completely by brute force. Claude Shannon

---

* The word 'algebra' is derived from al-Jabr, part of the title of a book by none other than al-Khwarizmi.

wrote the first-ever academic paper about computer chess in 1950. The problem, he explained, 'is not that of designing a machine to play perfect chess (which is quite impractical) nor one which merely plays legal chess (which is trivial). We would like to play a skilful game, perhaps comparable to that of a good human player.'[7]

'Comparable to that of a good human.' That is where this is all going, is it not? Could a mere algorithm, grinding mindlessly through a pre-specified procedure, outperform the human mind? And what else might such an algorithm achieve? That was why Turing and Shannon were intrigued by computer chess. It wasn't about the chessboard; it was about whether machines could think.

The theory that chess required thinking persisted for decades. In 1979, in *Gödel, Escher, Bach*, a book about the emergence of intelligence, Douglas Hofstadter argued that a computer that was sophisticated enough to be the world champion at chess simply couldn't help but be intelligent in other ways. 'I'm bored with chess,' it might reply when challenged to a game. 'Let's talk about poetry.'[8]

Hofstadter didn't dismiss the possibility that an algorithm could play chess brilliantly. He just thought that the algorithm would have to be so subtle, complex and multifaceted that chess would be the least of its achievements.

Hofstadter was wrong. Just 18 years later, a computer – IBM's Deep Blue – beat a human world champion, Kasparov himself. Deep Blue never broached the topic of poetry. It worked much like TuroChamp, except that it examined 150 million positions a second – looking not just two moves ahead, but many – and was supported by a huge library of human game openings. One move that dismayed – and defeated – Kasparov was simply plucked out of Deep Blue's library of games played by humans. How mundane.[9]

Deep Blue's victory showed that brute force calculation really could be a substitute for the mysterious qualities of the human mind. Maybe algorithms don't really need to 'think'. Maybe thinking doesn't matter as much as we assumed it did.

Hofstadter was exasperated by the narrowness of Deep Blue's capabilities. Without flexible, general human-like intelligence, he complained, it was 'trickery' to describe an algorithm as artificial intelligence, no matter how expert it might be at some particular task such as chess.[10] But IBM's approach with Deep Blue is typical of successful algorithmic design today: programmers are happy to borrow ideas from neuroscience (digital neural nets, for example, are inter-connected logic nodes that mimic some aspects of how an animal brain works), but they don't try to emulate human cognition. What matters is not insight into consciousness, but results – whether such results are 'trickery' or not. And these days, despite a continued lack of interest in discussing poetry, delivering results is what computers do.

Consider, for example, CloudCV, a system that accurately answers informally phrased answers about photographs. I tried it out on a photograph of some young people hang-ing out in someone's front room. 'What are they doing?' I typed. 'What are they drinking?' CloudCV promptly – and correctly – told me they were playing on a Wii and drink-ing beer.[11]

Algorithms such as CloudCV still aren't as good at humans at dealing with such a wide-ranging set of questions. But they're getting better – and we aren't. As recently as 2016, the machines scored 55 per cent on a standardised challenge for answering visual questions; humans score 81 per cent. But by the summer of 2019 the machines were at 75 per cent. If you are confident they won't overtake us, you are more confident than I am.[12] Few people remember that Kasparov solidly beat

Deep Blue in 1996. By the rematch in 1997, Deep Blue was twice as powerful, and Kasparov wasn't. Deep Blue would have been twice as powerful again in 1998. Kasparov's defeat was only a matter of time.

This isn't just going to happen on the chessboard. Already, algorithms – those step-by-step instructions – can diagnose skin cancer, breast cancer or diabetes at a level that is rapidly surpassing skilled medical professionals. These feats of pattern recognition are often accomplished by multiple layers of neural networks. They are not algorithms that Euclid, or even Alan Turing, would have recognised. But they are algorithms nonetheless.[13]

As it becomes clear that algorithms will achieve a performance 'comparable to that of a good human' in more and more fields, economists have increasingly pondered the implications for work. The received wisdom has been shaped by a 2003 article by David Autor, Frank Levy and Richard Murnane. (We met it already in the chapters on spreadsheets and chatbots.) They argued that most 'jobs' are bundles of 'tasks', some of which are routine and some of which are non-routine. Algorithms constantly encroach on routine tasks.[14] This distinction has proved a powerful way of understanding the implications of computers in the workplace: as computers take over tasks, jobs are more likely to change than disappear.

There is a problem, however: it is not always obvious where the routine ends and the non-routine begins. Who would describe diagnosing cancer as routine? But perhaps this should already have been obvious to us from the example of chess, which seemed so ineffable for so long.

Among the most impressive things algorithms can now do more skilfully than humans is ... write algorithms. AlphaZero is a game-learning algorithm developed by DeepMind, a sister company of Google. AlphaZero 'plays like a human on fire', says the former British chess champion Matthew Sadler. And it effectively programmed itself: humans wrote the learning algorithm, then the learning algorithm wrote a chess-playing algorithm.[15] In 2017 AlphaZero trained itself in a matter of a few hours to thrash the best chess-playing software, Stockfish, which easily beats the best humans. Stockfish examines 60 million positions a second; AlphaZero examines just 60,000. It wins anyway, because its neural network is simply better at recognising the patterns of the game.[16]

In the pages of this book we've seen many reasons not to worry when technology makes jobs obsolete, from the spreadsheet to the printing press to the sewing machine. What we don't know is if this time is different – because we don't know whether the very idea of a 'non-routine' task is starting to disappear. We are learning that step-by-step can take you a very long way when you don't, as Alan Turing did, need to plan each of those steps with a piece of paper and a humble pencil.

Notes

1 The Pencil

1 Henry David Thoreau, *The Maine Woods* (1864), https://en.wikisource.org/wiki/The_Maine_Woods_(1864)/Appendix#322.
2 Henry Petroski, *The Pencil: A History of Design and Circumstance* (London: Faber and Faber, 1989).
3 *Encyclopaedia Britannica* (1771 edition).
4 Petroski, *The Pencil*, p. 6.
5 Leonard Read, 'I, Pencil: My Family Tree as Told to Leonard E. Read', *The Freeman* (1958) available at https://fee.org/resources/i-pencil/.
6 Read, 'I, Pencil'.
7 Petroski, *The Pencil*, pp. 184–6.
8 https://pencils.com/pencil-making-today-2/.
9 Read, 'I, Pencil'.
10 Milton Friedman, *Free to Choose* (PBS, 1980), available at: https://www.youtube.com/watch?v=67tHtpac5ws.
11 https://geology.com/minerals/graphite.shtml.
12 Eric Voice, 'History of the Manufacture of Pencils', *Transactions of the Newcomen Society* 27 (1950).
13 Petroski, *The Pencil*, pp. 62, 69.
14 John Quiggin, 'I, Pencil, Product of the Mixed Economy', https://johnquiggin.com/2011/04/16/i-pencil-a-product-of-the-mixed-economy/; Newell Rubbermaid corporate website.

2 Bricks

1 Stefanie Pietkiewicz, 'From brick to marble: Did Augustus Caesar really transform Rome?', UCLA press release, 3 March 2015, http://newsroom.ucla.edu/stories/from-brick-to-marble:-did-augustus-caesar-really-transform-rome.
2 Hannah B. Higgins, *The Grid Book* (Cambridge, MA: MIT Press, 2009), p. 25.

3 Gavin Kennedy, https://www.adamsmith.org/blog/economics/of-pins-and-things.

4 James W. P. Campbell and Will Pryce, *Brick: A World History* (London: Thames & Hudson, 2003), p. 186.

5 New International Bible, https://www.biblegateway.com/passage/?search=Genesis+11%3A1-9&version=NIV.

6 Campbell and Pryce, *Brick: A World History*.

7 According to an anecdote by Alvar Aalto, who elaborated that 'Architecture is the transformation of a worthless brick into something worth its weight in gold.' http://uk.phaidon.com/agenda/architecture/articles/2015/april/01/even-modernists-like-mies-loved-bricks/.

8 Campbell and Pryce, *Brick: A World History*, pp. 26–7.

9 Campbell and Pryce, *Brick: A World History*, pp. 28–9.

10 Campbell and Pryce, *Brick: A World History*, p. 30.

11 http://www.world-housing.net/major-construction-types/adobe-introduction; Campbell and Pryce, *Brick: A World History*, p. 30.

12 Esther Duflo and Abhijit Banerjee, *Poor Economics* (New York: Public Affairs, 2011).

13 Edward Dobson and Alfred Searle, *Rudimentary Treatise on the Manufacture of Bricks and Tiles*, 14th edn (London: The Technical Press, 1936).

14 Jesus Diaz, 'Everything You Always Wanted to Know About Lego', https://gizmodo.com/5019797/everything-you-always-wanted-to-know-about-lego.

15 Stewart Brand, *How Buildings Learn: What Happens After They're Built* (New York: Viking Press, 1994), p. 123.

16 Campbell and Pryce, *Brick: A World History*, p. 296.

17 Campbell and Pryce, *Brick: A World History*, p. 267.

18 Carl Wilkinson, 'Bot the builder: the robot that will replace bricklayers', *Financial Times*, 23 February 2018, https://www.ft.com/content/db2b5d64-10e7-11e8-a765-993b2440bd73.

19 http://iopscience.iop.org/article/10.1088/1755-1315/140/1/012127/pdf.

20 Peter Smisek, 'A Short History of "Bricklaying Robots"', 17 October 2017, https://www.theb1m.com/video/a-short-history-of-bricklaying-robots.

3 The Factory

1 Justin Corfield, 'Lombe, John', in Kenneth E. Hendrickson III (ed.), *The Encyclopedia of the Industrial Revolution in World History* (Lanham, MD: Rowman & Littlefield, 2014), p. 568.

2 Joshua Freeman, *Behemoth: A History of the Factory and the Making of the Modern World* (London: WW Norton, 2018), pp. 1–8.

3 Our World In Data, https://ourworldindata.org/economic-growth#the-total-output-of-the-world-economy-over-the-last-two-thousand-years.

4 Adam Smith, *An Inquiry into the Nature and Causes of the Wealth of Nations* (1776), available at https://www.ibiblio.org/ml/libri/s/SmithA_WealthNations_p.pdf.

5   William Blake, *Milton a Poem*, http://www.blakearchive.org/
     search/?search=jerusalem c1804-1811.

6   https://picturethepast.org.uk/image-library/image-details/poster/
     DCAV000798/posterid/DCAV000798.html.

7   https://www.bl.uk/collection-items/the-life-and-adventures-of-michael-
     armstrong-the-factory-boy; Freeman, *Behemoth*, p. 25.

8   Bill Cahn, *Mill Town* (New York: Cameron and Kahn, 1954).

9   Friedrich Engels, *The Condition of the Working Class in England* (1845).

10   F. W. Taylor, *Principles of Scientific Management* (1911), p. 83, https://archive.
     org/stream/principlesofscie00taylrich#page/83/mode/2up.

11   Daniel A. Wren and Arthur G. Bedeian, 'The Taylorization of Lenin:
     Rhetoric or Reality', *International Journal of Social Economics* 31.3 (2004),
     cited in Freeman, *Behemoth*, pp. 174–5. See also Stephen Kotkin, *Magnetic
     Mountain* (Chicago, IL: University of Chicago Press, 1997); Loren Graham,
     The Ghost of the Executed Engineer (Cambridge, MA: Harvard University
     Press, 1996).

12   Note that exploitative conditions are still found in some rich countries.
     See, for instance, Sarah O'Connor, 'Dark Factories', *Financial Times*,
     17 May 2018, https://www.ft.com/content/e427327e-5892-11e8-b8b
     2-d6ceb45fa9d0.

13   Myself included: Tim Harford, *The Undercover Economist* (New York: Oxford
     University Press, 2005). For a detailed account of conditions for women
     in Chinese factories see Pun Ngai, *Made in China* (Hong Kong: Hong
     Kong University Press, 2005). An intriguing study of Ethiopian factories is
     summarised by Christopher Blattman and Stefan Dercon, 'Everything We
     Knew About Sweatshops Was Wrong', *New York Times*, 27 April 2017.

14   Freeman, *Behemoth*, p. 8; Elizabeth Roberts, *Women's Work 1840–1940*
     (Basingstoke: Macmillan Education, 1988).

15   Wolfgang Streeck, 'Through Unending Halls', *London Review of Books* 41.3
     (7 February 2019), pp. 29–31, https://www.lrb.co.uk/the-paper/v41/n03/
     wolfgang-streeck/through-unending-halls.

16   For the lower figure, see https://www.nytimes.com/2012/01/22/business/
     apple-america-and-a-squeezed-middle-class.html, and for the higher figure,
     see http://focustaiwan.tw/news/aeco/201008190012.aspx.

17   Niall McCarthy, 'The World's Biggest Employers Infographic',
     Forbes, with data from Statista, https://www.forbes.com/sites/
     niallmccarthy/2015/06/23/the-worlds-biggest-employers-infographic/#7
     087ca18686b.

18   Charles Duhigg and Keith Bradsher, 'How the U.S. Lost Out on
     iPhone Work', *New York Times*, 21 January 2012, https://www.nytimes.
     com/2012/01/22/business/apple-america-and-a-squeezed-middle-class.
     html; 'Light and Death', *Economist*, 27 May 2010, https://www.economist.
     com/business/2010/05/27/light-and-death.

19   Leslie T. Chang, 'The Voices of China's Workers', TED Talk, 2012, https://
     www.ted.com/talks/leslie_t_chang_the_voices_of_china_s_workers/
     transcript?language=en.

20 China Labour Bulletin Strike Map, https://clb.org.hk/content/introduction-china-labour-bulletin%E2%80%99s-strike-map.

21 Yan Yuang, 'Inside China's Crackdown on Young Marxists', *Financial Times Magazine*, 14 February 2019, https://www.ft.com/content/fd087484-2f23-11e9-8744-e7016697f225.

22 James Fallows, 'Mr. China Comes to America', *Atlantic*, December 2012, https://www.theatlantic.com/magazine/archive/2012/12/mr-china-comes-to-america/309160/.

23 Quoted in Adam Menuge, 'The Cotton Mills of the Derbyshire Derwent and its Tributaries', *Industrial Archaeology Review* 16.1 (1993), 38–61, DOI: 10.1179/iar.1993.16.1.38; see also Neil Cossons, *Making of the Modern World: Milestones of Science and Technology* (London: John Murray, 1992).

24 Fallows, 'Mr. China Comes to America'.

25 Charles Babbage, *On the Economy of Machinery and Manufactures* (London: Charles Knight, 1832; reprinted Cambridge: Cambridge University Press, 2009).

26 Richard Baldwin, 'Globalisation, automation and the history of work: Looking back to understand the future', 31 January 2019, https://voxeu.org/content/globalisation-automation-and-history-work-looking-back-understand-and-future; Richard Baldwin, *The Great Convergence* (Cambridge, MA: Harvard University Press, 2016).

27 BetaNews, 'The Global Supply Chain Behind the iPhone 6', https://betanews.com/2014/09/23/the-global-supply-chain-behind-the-iphone-6/.

4 The Postage Stamp

1   Rowland Hill, *Post Office Reform: Its Importance and Practicability*, 3rd edition (1837), available at http://www.gbps.org.uk/information/downloads/files/penny-postage/Post%20Office%20Reform,%20its%20Importance%20and%20Practicability%20-%20Rowland%20Hill%20(3rd%20edition,%201837).pdf, p. iv.

2   Sir Rowland Hill and George Birkbeck Hill, *The Life of Sir Rowland Hill and the History of Penny Postage* (1880), available at http://www.gbps.org.uk/information/downloads/files/penny-postage/The%20Life%20of%20Sir%20Rowland%20Hill%20(Volume%201).pdf, p. 263.

3   Hill and Hill, *Life of Sir Rowland Hill*, pp. 279, 326.

4   Hill, *Post Office Reform*.

5   Hill and Hill, *Life of Sir Rowland Hill*, p. 278.

6   Hill, *Post Office Reform*, p. 54.

7   Hill and Hill, *Life of Sir Rowland Hill*, pp. 364–371.

8   Hill and Hill, *Life of Sir Rowland Hill*, p. 238.

9   Gregory Clark, *Average Earnings and Retail Prices, UK, 1209–2017* (Davis, CA: University of California, Davis, 28 April 2018), https://www.measuringworth.com/datasets/ukearncpi/earnstudyx.pdf.

10 Hill and Hill, *Life of Sir Rowland Hill*, p. 238.

11 James Vernon, *Distant Strangers: How Britain Became Modern* (Berkeley: University of California Press, 2014), p. 68.

12 Hill, *Post Office Reform*, p. 80.

13 Hill, *Post Office Reform*, p. 68–81.

14 https://www.richmondfed.org/~/media/richmondfedorg/publications/research/economic_review/1992/pdf/er780201.pdf.

15 Hill, *Post Office Reform*, p. 79.

16 Hill, *Post Office Reform*, p. 79.

17 Eunice and Ron Shanahan, 'The Penny Post', The Victorian Web, http://www.victorianweb.org/history/pennypos.html.

18 Catherine J. Golden, *Posting It: The Victorian Revolution in Letter Writing* (Gainesville: University Press of Florida, 2009).

19 Randal Stross, 'The Birth of Cheap Communication (and Junk Mail)', *New York Times*, 20 February 2010, https://www.nytimes.com/2010/02/21/business/21digi.html.

20 Daron Acemoğlu, Jacob Moscona and James Robinson, 'State capacity and American technology: Evidence from the 19th century', *Vox*, 27 June 2016, https://voxeu.org/article/state-capacity-and-us-technical-progress-19th-century.

21 'Amazon is not the only threat to legacy post offices', *Economist*, 19 April 2018, https://www.economist.com/business/2018/04/19/amazon-is-not-the-only-threat-to-legacy-post-offices.

22 'The Shocking Truth about How Many Emails Are Sent', Campaign Monitor, March 2019, https://www.campaignmonitor.com/blog/email-marketing/2018/03/shocking-truth-about-how-many-emails-are-sent/.

23 Acemoğlu, Moscona and Robinson, 'State capacity and American technology'.

5 Bicycles

1 Margaret Guroff, *The Mechanical Horse: How the Bicycle Reshaped American Life* (Austin: University of Texas Press, 2016), ch. 1.

2 Harry Oosterhuis, 'Cycling, modernity and national culture', *Social History* 41.3 (2016), 233–48.

3 Guroff, *Mechanical Horse*, ch. 3.

4 Paul Smethurst, *The Bicycle – Towards a Global History* (London: Palgrave Macmillan, 2015), ch. 1.

5 David Herlihy, *Bicycle: The History* (New Haven, CT: Yale University Press, 2004), pp. 268–9.

6 Margaret Guroff, 'The Wheel, the Woman and the Human Body', https://longreads.com/2018/07/06/the-wheel-the-woman-and-the-human-body/ (extract from *The Mechanical Horse*).

7 Guroff, 'The Wheel, the Woman and the Human Body'.

8 Guroff, *Mechanical Horse*, ch. 3.

9 Karthik Muralidharan and Nishith Prakash, 'Cycling to School: Increasing Secondary School Enrollment for Girls in India', NBER Working Paper No. 19305, August 2013.

10 Jason Gay, 'The LeBron James interview about bicycles', *Wall Street Journal*, 6 August 2018, https://www.wsj.com/articles/the-lebron-james-interview-about-bicycles-1533561787.

11 William Manners, 'The secret history of 19th century cyclists', *Guardian*, 9 June 2015, https://www.theguardian.com/environment/bike-blog/2015/jun/09/feminism-escape-widneing-gene-pools-secret-history-of-19th-century-cyclists; Steve Jones, 'Steve Jones on Extinction: A conversation with Steve Jones', https://www.edge.org/conversation/steve_jones-steve-jones-on-extinction.

12 David A. Hounshell, *From the American System to Mass Production, 1800–1932* (Baltimore, MD: Johns Hopkins University Press, 1984), Introduction and ch. 5.

13 Jane Jacobs, *Cities and the Wealth of Nations* (New York: Random House, 1984), p. 150.

14 Smethurst, *The Bicycle*, ch. 1.

15 Jacobs, *Cities and the Wealth of Nations*, p. 38.

16 Smethurst, *The Bicycle*, ch. 3, p. 118.

17 Tatsuzo Ueda, 'The development of the bicycle industry in Japan after World War II', *Japanese Experience of the UNU Human and Social Development Programme* (1981), https://d-arch.ide.go.jp/je_archive/english/society/wp_je_unu38.html.

18 World Bank Blog, 'Cycling Is Everyone's Business', https://blogs.worldbank.org/publicsphere/cycling-everyone-s-business; Worldometers: Bicycles, http://www.worldometers.info/bicycles/.

19 The European Cyclists Federation reports 16 million shared bikes in China alone in 2018: https://ecf.com/news-and-events/news/executive-summary-what-happening-bike-share-world-1; *The Economist* counted more than 1500 bike-share schemes in 2017, with more coming at an increasing rate: https://www.economist.com/christmas-specials/2017/12/19/how-bike-sharing-conquered-the-world.

6 Spectacles

1 Thomas Black, 'Google Glass Finds a New Home at the Factory', *Bloomberg*, 20 May 2019, https://www.bloomberg.com/news/articles/2019-05-20/google-glass-finds-a-new-home-at-the-factory.

2 Andre Bourque, 'Smart glasses are making workers more productive', CIO, 16 May 2017, https://www.cio.com/article/3196294/smart-glasses-are-making-workers-more-productive.html.

3 Project Glass: Live Demo at Google I/O, 27 June 2012, https://www.youtube.com/watch?v=D7TB8b2t3QE.

4 Black, 'Google Glass Finds a New Home at the Factory'.

5 https://en.wikipedia.org/wiki/Ibn_al-Haytham.

6 David C. Lindberg, *Theories of Vision from al-Kindi to Kepler* (Chicago, IL: University of Chicago Press, 1981), pp. 209–10.

7 Rebecca Stefoff, *Microscopes and Telescopes* (New York: Marshall Cavendish, 2007), pp. 12–13.

8 Lindberg, *Theories of Vision*, p. 86.

9 Steven Johnson, *How We Got to Now* (New York: Penguin, 2014), pp. 15–16.

10   James B. Tschen-Emmons, *Artifacts from Medieval Europe* (Santa Barbara, CA: ABC-CLIO, 2015), p. 260.

11   Alberto Manguel, *A History of Reading* (London: Flamingo, 1997), p. 293.

12   Manguel, *History of Reading*, p. 292.

13   Born circa 1255, according to https://en.wikipedia.org/wiki/Jordan_of_Pisa.

14   Steven R. Fischer, *A History of Reading* (London: Reaktion Books, 2004), p. 186.

15   Stefoff, *Microscopes and Telescopes*, pp. 14–16.

16   Stefoff, *Microscopes and Telescopes*, pp. 14–16.

17   'Britain's Eye Health in Focus: A snapshot of consumer attitudes and behaviour towards eye health' (College of Optometrists, 2013), http://www.wcb-ccd.org.uk/perspectif/library/BEH_Report_FINAL%20(1).pdf.

18   'VisionWatch' (Vision Council, September 2016), https://www.thevisioncouncil.org/sites/default/files/research/VisionWatch_VisionCouncil_Member_Benefit_Report_September%202016_FINAL.pdf; 'Share of Japanese wearing eyeglasses as of September 2017, by age group and gender', Statista, https://www.statista.com/statistics/825746/japan-glasses-usage-share-by-age-gender/.

19   'Eyeglasses for Global Development: Bridging the Visual Divide' (Geneva: World Economic Forum, June 2016), http://www3.weforum.org/docs/WEF_2016_EYElliance.pdf.

20   Sam Knight, 'The spectacular power of Big Lens', *Guardian*, 10 May 2018, https://www.theguardian.com/news/2018/may/10/the-invisible-power-of-big-glasses-eyewear-industry-essilor-luxottica.

21   'Eyeglasses for Global Development'.

22   Priya Adhisesha Reddy et al., 'Effect of providing near glasses on productivity among rural Indian tea workers with presbyopia (PROSPER): a randomised trial', *Lancet Global Health* 6.9, PE1019-E1027 (1 September 2018), http://dx.doi.org/10.1016/ S2214-109X(18)30329-2.

23   'Eyeglasses to Improve Workers' Manual Dexterity', Givewell, April 2019, https://www.givewell.org/international/technical/programs/eyeglasses-workers-manual-dexterity.

24   'Eyeglasses for Global Development'.

25   Paul Glewwe, Albert Park, Meng Zhao, *A Better Vision for Development: Eyeglasses and Academic Performance in Rural Primary Schools in China*, HKUST IEMS Working Paper No. 2015-37, June 2016, https://www.povertyactionlab.org/sites/default/files/publications/424_542_A%20better%20vision%20for%20development_PaulGlewwe_May2016.pdf.

26   Elie Dolgin, 'The myopia boom', *Nature* 519.7543 (18 March 2015), https://www.nature.com/news/the-myopia-boom-1.17120?WT.mc_id=TWT_NatureNews#/eye.

27   John Trevelyan and Peter Ackland, 'Global Action Plan Indicators – the data in full', Vision Atlas, International Agency for the Prevention of Blindness, updated 11 October 2018, http://atlas.iapb.org/global-action-plan/gap-indicators/.

28   Jennifer L. Y. Yip et al., 'Process evaluation of a National Primary Eye
      Care Programme in Rwanda', *BMC Health Services Research* 18.1 (December
      2018), https://doi.org/10.1186/s12913-018-3718-1.
29   Zhang et al., 'Self correction of refractive error among young people in
      rural China: results of cross sectional investigation', *BMJ* 2011;343:d4767,
      doi: 10.1136/bmj.d4767, http://cvdw.org/resources/bmj.d4767.full.pdf.

7 Canned Food

1   Alex Davies, 'Inside the Races That Jump-Started the Self-Driving Car',
      Wired, 11 October 2017, https://www.wired.com/story/darpa-grand-urba
      n-challenge-self-driving-car/.
2   Alex Davies, 'Inside the Races'; 'An Oral History of the Darpa Grand
      Challenge, the Grueling Robot Race That Launched the Self-Driving Car',
      Wired, 3 August 2017, https://www.wired.com/story/darpa-grand-challenge-
      2004-oral-history/.
3   https://en.wikipedia.org/wiki/History_of_self-driving_cars.
4   *Inventors and Inventions* (New York: Marshall Cavendish, 2008).
5   Kat Eschner, 'The Father of Canning Knew His Process Worked, But Not
      Why It Worked', *Smithsonian* magazine, 2 February 2017, https://www.
      smithsonianmag.com/smart-news/father-canning-knew-his-process-
      worked-not-why-it-worked-180961960/.
6   http://www.oxfordreference.com/view/10.1093/oi/
      authority.20110803095425331.
7   *Inventors and Inventions*.
8   N. Appert, *The Art of Preserving All Kinds of Animal and Vegetable
      Substances for Several Years* (1812), available at http://www.gutenberg.org/
      files/52551/52551-h/52551-h.htm.
9   Tom Geoghegan, 'The story of how the tin can nearly wasn't', BBC
      News Magazine, 21 April 2013, https://www.bbc.co.uk/news/
      magazine-21689069.
10   Geoghegan, 'The story'.
11   Vivek Wadhwa, 'Silicon Valley Can't Be Copied', *MIT Technology
      Review*, 3 July 2013, https://www.technologyreview.com/s/516506/
      silicon-valley-cant-be-copied/.
12   https://en.wikipedia.org/wiki/Category:Information_technology_places.
13   Wadhwa, 'Silicon Valley'.
14   Geoghegan, 'The story'.
15   Sue Shephard, *Pickled, Potted, and Canned: How the Art and Science of Food
      Preserving Changed the World* (New York: Simon and Schuster, 2006).
16   Geoghegan, 'The story'.
17   Zeynep Tufekci, 'How social media took us from Tahrir Square to
      Donald Trump', *MIT Technology Review*, 14 August 2018, https://www.
      technologyreview.com/s/611806/how-social-media-took-us-from-tahrir-
      square-to-donald-trump/.
18   Evan Osnos, 'Doomsday Prep for the Super-Rich', *New Yorker*, 22

January 2017, https://www.newyorker.com/magazine/2017/01/30/
doomsday-prep-for-the-super-rich.

19 *Inventors and Inventions*.

8 Auctions

1 Edward Gibbon, *The History of the Decline and Fall of the Roman Empire*
(1776–89), ch. 31, https://ebooks.adelaide.edu.au/g/gibbon/edward/g43d/
chapter31.html.

2 Herodotus, *The Histories* (Harmondsworth: Penguin Books, 1972), pp. 120–1.

3 Ralph Cassady, *Auctions and Auctioneering* (Berkeley: University of California
Press, 1967), pp. 33–6.

4 Brian Learmount, *The History of the Auction* (London: Barnard and
Learmount, 1985), p. 84.

5 Learmount, *History of the Auction*, p. 84.

6 Samuel Pepys, *Diary*, 6 November 1660, https://www.pepysdiary.com/
diary/1660/11/06/, 3 September 1662, https://www.pepysdiary.com/
diary/1662/09/03/.

7 John McMillan, *Reinventing the Bazaar: A Natural History of Markets* (New
York: WW Norton, 2002); 'Aalsmeer Flower Auction Fights the Clock',
New York Times video, 23 December 2014, https://www.youtube.com/
watch?v=zx7buFdpis4.

8 William Vickrey, 'Counterspeculation, Auctions, and Competitive Sealed
Tenders', *Journal of Finance* 16.1: 8–39.

9 Paul Klemperer, 'What Really Matters in Auction Design', *Journal of
Economic Perspectives* 16.1: 169–89, DOI: 10.1257/0895330027166.

10 Ken Binmore and Paul Klemperer, 'The Biggest Auction Ever: The Sale of
the British 3G Telecom Licenses', *Economic Journal* 112.478 (March 2002):
C74–C96, https://doi.org/10.1111/1468-0297.00020.

11 Google has a video tutorial explaining the workings of the Ad Auction:
https://www.youtube.com/watch?v=L5r0Ng8XbDs. For a more academic
discussion, Google's Chief Economist Hal Varian published an analysis of
the auction as it was in 2009. Hal Varian, 'Position Auctions', *International
Journal of Industrial Organization* 25.6 (2007): 1163–78.

12 Rachel Lerman, 'Google reports $7.1 billion profit, but still falls short on
third-quarter expectations', Associated Press, 28 October 2019.

13 Jasmine Enberg, 'Global Digital Ad Spending 2019', *eMarketer*, 28 March
2019, https://www.emarketer.com/content/global-digital-ad-
spending-2019.

14 Jack Nicas, 'Google Uses Its Search Engine to Hawk Its Products', *Wall
Street Journal*, 19 January 2017.

9 Tulips

1 Charles Mackay, *Extraordinary Popular Delusions and the Madness of
Crowds* (1841).

2   Mike Dash, *Tulipomania* (London: Phoenix, 2003).

3   Anne Goldgar, *Tulipmania: Money, Honor, and Knowledge in the Dutch Golden Age* (Chicago: University of Chicago Press, 2007).

4   Goldgar, *Tulipmania*.

5   Goldgar, *Tulipmania*.

6   Dash, *Tulipomania*.

7   Dash, *Tulipomania*.

8   Stephen Moss, 'The Super-studs: Inside the Secretive World of Racehorse Breeding', *Guardian*, 28 October 2009, https://www.theguardian.com/sport/2009/oct/28/sea-the-stars-stud.

9   Peter Garber, 'Famous First Bubbles', *Journal of Economic Perspectives*, Spring 1990; 'Tulipmania', *Journal of Political Economy* 97.3 (June 1989): 535–60.

10 James McClure and David Chandler Thomas, 'Explaining the timing of tulipmania's boom and bust: historical context, sequester capital and market signals', *Financial History Review*, 2017.

11 Andrew Odlyzko, 'Collective hallucinations and inefficient markets: The British Railway Mania of the 1840s', Working Paper, School of Mathematics and Digital Technology Center, University of Minnesota, 2010.

10 Queen's Ware

1   Brian Dolan, *Josiah Wedgwood: Entrepreneur to the Enlightenment* (London: Harper Perennial, 2004), p. 169.

2   Dolan, *Josiah Wedgwood*, p. 169.

3   Dolan, *Josiah Wedgwood*, p. 153.

4   Dolan, *Josiah Wedgwood*, pp. 213–14.

5   Katie Hafner and Brad Stone, 'IPhone Owners Crying Foul Over Price Cut', *New York Times*, 7 September 2007, https://www.nytimes.com/2007/09/07/technology/07apple.html.

6   R. H. Coase, 'Durability and Monopoly', *Journal of Law and Economics* 15.1 (April 1972): 143–9.

7   Nancy F. Koehn, *Brand New: How Entrepreneurs Earned Consumers' Trust from Wedgwood to Dell* (Boston, MA: Harvard Business School Press, 2001), p. 40.

8   Dolan, *Josiah Wedgwood*, p. 277.

9   See e.g. Emile Durkheim, *La science positive de la morale en Allemagne* (1887), available at https://gallica.bnf.fr/ark:/12148/bpt6k171631/f61. image; Thorstein Veblen, *The Theory of the Leisure Class* (1899); https://en.wikipedia.org/wiki/Trickle-down_effect.

10 Deniz Atik & A. Fuat Fırat, 'Fashion creation and diffusion: The institution of marketing', *Journal of Marketing Management* 29.7–8 (2013): 836–60, DOI: 10.1080/0267257X.2012.729073.

11 Malcolm Gladwell, 'The Coolhunt', *New Yorker*, 10 March 1997, https://www.newyorker.com/magazine/1997/03/17/the-coolhunt.

12 Vanessa Grigoriadis, 'Slaves of the Red Carpet', *Vanity Fair*, 10 February 2014, https://www.vanityfair.com/hollywood/2014/03/hollywood-fashion-stylists-rachel-zoe-leslie-fremar.

13   Nancy F. Koehn, *Brand New: How Entrepreneurs Earned Consumers' Trust from Wedgwood to Dell* (Boston, MA: Harvard Business School Press, 2001), p. 35.
14   Koehn, *Brand New*, p. 12.
15   Dolan, *Josiah Wedgwood*, pp. 174–5.
16   Dolan, *Josiah Wedgwood*, p. 263.
17   Dolan, *Josiah Wedgwood*, p. 217.
18   Dolan, *Josiah Wedgwood*, p. 287.
19   Jenny Uglow, *The Lunar Men: The Friends Who Made the Future* (London: Faber and Faber, 2002), p. 205.
20   See e.g. Wolfgang Pesendorfer, 'Design Innovation and Fashion Cycles', *American Economic Review* 85.4 (1995): 771–92; Barak Y. Orbach, 'The Durapolist Puzzle: Monopoly Power in Durable-Goods Markets', *Yale Journal on Regulation* 21.1 (2004): 67–119.
21   Dolan, *Josiah Wedgwood*, p. 277.

11 The Bonsack Machine

1   Allan M. Brandt, *The Cigarette Century: The Rise, Fall, and Deadly Persistence of the Product That Defined America* (New York: Basic Books, 2007).
2   Robert Proctor, *Golden Holocaust: Origins of the Cigarette Catastrophe and the Case for Abolition* (Berkeley: University of California Press, 2011).
3   Brandt, *Cigarette Century*.
4   Proctor, *Golden Holocaust*.
5   Proctor, *Golden Holocaust*.
6   Brandt, *Cigarette Century*.
7   Brandt, *Cigarette Century*.
8   Brandt, *Cigarette Century*.
9   Proctor, *Golden Holocaust*.
10   Terrence H. Witkowski, 'Promise Them Everything: A Cultural History of Cigarette Advertising Health Claims', *Current Issues and Research in Advertising* 13.2 (1991): 393–409.
11   Witkowski, 'Promise Them Everything'.
12   Brandt, *Cigarette Century*.
13   Witkowski, 'Promise Them Everything'.
14   'Smoke Gets In Your Eyes', *Mad Men* pilot episode, first broadcast 19 July 2007. Writer: Matthew Weiner. Director: Alan Taylor. Production company: Lionsgate. A video clip and transcript of this particular moment is at http://www.sarahvogelsong.com/blog/2018/1/29/its-toasted.
15   Proctor, *Golden Holocaust*.
16   Brandt, *Cigarette Century*.
17   https://en.wikipedia.org/wiki/Regulation_of_nicotine_marketing.
18   'Plain packaging of tobacco products: Evidence, design and implementation' (World Health Organization, 2016), https://www.who.int/tobacco/publications/industry/plain-packaging-tobacco-products/en/.
19   See e.g. William Savedoff, 'Plain packaging', British American Tobacco, http://www.bat.com/plainpackaging; 'Tobacco Companies Fail the

Corporate Social Responsibility Test of a Free-Market Advocate', Center for Global Development, 17 August 2017, https://www.cgdev.org/blog/tobacco-companies-fail-corporate-social-responsibility-test-free-market-advocate.

20　See e.g. Brandt, *The Cigarette Century*.

21　Global Health Observatory, World Health Organization, https://www.who.int/gho/tobacco/use/en/.

22　Proctor, *Golden Holocaust*.

23　The population just over doubled according to http://www.chinatoday.com/data/china.population.htm. Cigarette consumption went up around 20-fold according to Proctor, *Golden Holocaust*.

24　Population figures from http://www.chinatoday.com/data/china.population.htm. Cigarette consumption figures from Proctor, *Golden Holocaust*.

25　S. S. Xu, S. Gravely, G. Meng et al., 'Impact of China National Tobacco Company's "Premiumization" Strategy: longitudinal findings from the ITC China Surveys (2006–2015)', *Tobacco Control* 28.suppl 1 (29 August 2018), https://tobaccocontrol.bmj.com/content/early/2018/08/27/tobaccocontrol-2017-054193.

26　Xu et al., '"Premiumization" Strategy'.

12　The Sewing Machine

1　https://www.youtube.com/watch?v=koPmuEyP3a0.

2　https://budpride.co.uk/.

3　See e.g. Rachel Alexander, 'Woke Capitalism: Big Business Pushing Social Justice Issues', The Stream, 12 June 2019, https://stream.org/woke-capitalism-big-business-pushing-social-justice-issues/.

4　https://en.wikipedia.org/wiki/Declaration_of_Sentiments (accessed 8 July 2019).

5　Ruth Brandon, *Singer and the Sewing Machine: A Capitalist Romance* (London: Barrie & Jenkins, 1977), p. 42.

6　Adam Mossoff, 'The Rise and Fall of the First American Patent Thicket: The Sewing Machine War of the 1850s', *Arizona Law Review* 53 (2011): 165–211.

7　'The Story of the Sewing-Machine; Its Invention Improvements Social, Industrial and Commercial Importance', *New York Times*, 7 January 1860, https://www.nytimes.com/1860/01/07/archives/the-story-of-the-sewingmachine-its-invention-improvements-social.html.

8　*Godey's Lady's Book and Magazine* 61 (1860), p. 77.

9　Brandon, *Singer*, p. 44.

10　Brandon, *Singer*, p. 21.

11　Brandon, *Singer*, p. 45.

12　'Story of the Sewing-Machine'.

13　Grace Rogers Cooper, *The Sewing Machine: Its Invention and Development* (Washington, DC: Smithsonian Institution, 1968), pp. 41–2.

14　Mossoff, 'American Patent Thicket'.

15 Mossoff, 'American Patent Thicket'.
16 'Patent pools and antitrust – a comparative analysis', Secretariat, World Intellectual Property Organization, March 2014, https://www.wipo.int/export/sites/www/ip-competition/en/studies/patent_pools_report.pdf.
17 David A. Hounshell, *From the American System to Mass Production, 1800–1932* (Baltimore, MD: Johns Hopkins University Press, 1984), ch. 2.
18 Mossoff, 'American Patent Thicket'.
19 Mossoff, 'American Patent Thicket'.
20 Brandon, *Singer*, p. 117.
21 Andrew Godley, 'Selling the Sewing Machine Around the World: Singer's International Marketing Strategies, 1850–1920', *Enterprise and Society* 7.2 (March 2006).
22 Brandon, *Singer*, p. 140.
23 Brandon, *Singer*, pp. 120–1.
24 Mossoff, 'American Patent Thicket'.
25 Brandon, *Singer*, p. 125.
26 Brandon, *Singer*, pp. 68, 73.
27 Brandon, *Singer*, p. 124.
28 Brandon, *Singer*, p. 127.
29 'Story of the Sewing-Machine'.
30 *Godey's Lady's Book and Magazine* 61, p. 77.
31 Mossoff, 'American Patent Thicket'.

13 The Mail-Order Catalogue

 1 https://www.wards.com/.
 2 David Blanke, *Sowing the American Dream: How Consumer Culture Took Root in the Rural Midwest* (Athens: Ohio University Press, 2000).
 3 Blanke, *American Dream*.
 4 Blanke, *American Dream*.
 5 Doug Gelbert, *Branded! Names So Famous the People Have Been Forgotten* (Cruden Bay Books, 2016).
 6 Gelbert, *Branded!*.
 7 Blanke, *American Dream*.
 8 Gelbert, *Branded!*.
 9 Gelbert, *Branded!*.
10 Blanke, *American Dream*.
11 Leslie Kaufman with Claudia H. Deutsch, 'Montgomery Ward to Close Its Doors', *New York Times*, 29 December 2000, https://www.nytimes.com/2000/12/29/business/montgomery-ward-to-close-its-doors.html.
12 'Ward (Montgomery) & Co.', Encyclopaedia of Chicago, http://www.encyclopedia.chicagohistory.org/pages/2895.html.
13 Catalogue No. 13, Montgomery Ward & Co., available at https://archive.org/details/catalogueno13spr00mont.
14 Earle F. Walbridge, '*One Hundred Influential American Books Printed before 1900. Catalogue and Addresses. Exhibition at The Grolier Club*

April Eighteenth–June Sixteenth MCMXLVI' (review), *The Papers of the Bibliographical Society of America* 41.4 (Fourth Quarter, 1947): 365–7.

15 Blanke, *American Dream*.

16 *99 Percent Invisible*, 'The House that Came in the Mail', 11 September 2018, https://99percentinvisible.org/episode/the-house-that-came-in-the-mail/.

17 According to the Measuring Worth website, $30m in 1900 is worth $0.9bn at 2018 prices, or $4.2bn if adjusted in line with unskilled wages.

18 Judith M. Littlejohn, 'The Political, Socioeconomic, and Cultural Impact of the Implementation of Rural Free Delivery in Late 1890s US' (2013). https://digitalcommons.brockport.edu/cgi/viewcontent.cgi?article=1009&context=hst_theses.

19 Littlejohn, 'Political, Socioeconomic, and Cultural Impact'.

20 Littlejohn, 'Political, Socioeconomic, and Cultural Impact'.

21 Montgomery Ward Catalogue extract, 1916, Carnival Glass Worldwide, https://www.carnivalglassworldwide.com/montgomery-ward-ad-1916.html.

22 1911 Modern Homes Catalog, available at http://www.arts-crafts.com/archive/sears/page11.html.

23 Nancy Keates, 'The Million-Dollar Vintage Kit Homes That Come From Sears', *Wall Street Journal*, 21 September 2017, https://www.wsj.com/articles/some-vintage-kit-homes-now-sell-for-over-1-million-1506001728.

24 'Montgomery Ward to End Catalogue Sales', *Los Angeles Times*, 2 August 1985, https://www.latimes.com/archives/la-xpm-1985-08-02-mn-5529-story.html.

25 Stephanie Strom, 'Sears Eliminating Its Catalogues and 50,000 Jobs', *New York Times*, 26 January 1993, https://www.nytimes.com/1993/01/26/business/sears-eliminating-its-catalogues-and-50000-jobs.html.

26 Larry Riggs, 'Direct Mail Gets Most Response, But Email Has Highest ROI: DMA', *Chief Marketer*, 22 June 2012, https://www.chiefmarketer.com/direct-mail-gets-most-response-but-email-has-highest-roi-dma/.

27 Jiayang Fan, 'How e-commerce is transforming rural China', *New Yorker*, 16 July 2018, https://www.newyorker.com/magazine/2018/07/23/how-e-commerce-is-transforming-rural-china.

28 Anderlini, 'Liu Qiangdong, the "Jeff Bezos of China", on making billions with JD.com', *Financial Times*, 15 September 2017, https://www.ft.com/content/a257956e-97c2-11e7-a652-cde3f882dd7b.

29 Feng Hao, 'Will "Taobao villages" spur a rural revolution?', China Dialogue, 24 May 2016, https://www.chinadialogue.net/article/show/single/en/8943-Will-Taobao-villages-spur-a-rural-revolution.

30 Anderlini, 'Liu Qiangdong'.

31 James J. Feigenbaum and Martin Rotemberg, 'Communication and Manufacturing: Evidence from the Expansion of Postal Services', working paper, https://scholar.harvard.edu/files/feigenbaum_and_rotemberg_-_rural_free_delivery.pdf.

32 See e.g. Feng Hao, 'Will "Taobao villages" spur a rural revolution?'; Josh

Freedman, 'Once poverty-stricken, China's "Taobao villages" have found a lifeline making trinkets for the internet', Quartz, 12 February 2017, https://qz.com/899922/once-poverty-stricken-chinas-taobao-villages-have-found-a-lifeline-making-trinkets-for-the-internet/.

14 Fast-Food Franchises

1 Ray Kroc with Robert Anderson, *Grinding It Out: The Making of McDonald's* (Chicago, IL: Contemporary Books, 1987), ch 1.

2 John F. Love, *McDonald's: Behind the Arches* (London: Transworld, 1987), p. 23.

3 Kroc, *Grinding It Out*, 'Afterword'.

4 Love, *McDonald's*, pp. 16–19.

5 Love, *McDonald's*, pp. 25–6.

6 Love, *McDonald's*, p. 16.

7 Love, *McDonald's*, p. 24.

8 Love, *McDonald's*, p. 22.

9 Kroc, *Grinding It Out*, ch. 2.

10 Kroc, *Grinding It Out*, p. 176.

11 Roger D. Blair, Francine Lafontaine, *The Economics of Franchising* (Cambridge: Cambridge University Press, 2005).

12 Examples drawn from https://www.franchisedirect.com/top100globalfranchises/rankings.

13 Blair, Lafontaine, *Economics of Franchising*; Jaimie Seaton, 'Martha Matilda Harper, The Greatest Businesswoman You've Never Heard Of', Atlas Obscura, 11 January 2017, https://www.atlasobscura.com/articles/martha-matilda-harper-the-greatest-businesswoman-youve-never-heard-of.

14 Love, *McDonald's*, p. 53.

15 Love, *McDonald's*, ch. 3.

16 Love, *McDonald's*, pp. 148–9.

17 Love, *McDonald's*, pp. 144–6.

18 Hayley Peterson, 'Here's what it costs to open a McDonald's restaurant', Business Insider, 6 May 2019, https://www.businessinsider.com/what-it-costs-to-open-a-mcdonalds-2014-11.

19 https://en.wikipedia.org/wiki/McDonald's.

20 Peterson, 'Here's what it costs'.

21 Alan B. Kreuger, 'Ownership, Agency, and Wages: An Examination of Franchising in the Fast Food Industry', *Quarterly Journal of Economics* 106.1 (February 1991): 75–101.

22 Sugato Bhattacharyya and Francine Lafontaine, 'Double-Sided Moral Hazard and the Nature of Share Contracts', *RAND Journal of Economics* 26.4 (Winter 1995): 761–81.

23 Love, *McDonald's*, p. 24.

15 Fundraising Appeals

1 Adam Smith, *An Inquiry into the Nature and Causes of the Wealth of Nations* (1776).

2 'Gross Domestic Philanthropy: An international analysis of GDP, tax and giving', Charities Aid Foundation, January 2016, https://www.cafonline. org/docs/default-source/about-us-policy-and-campaigns/gross-domestic-philanthropy-feb-2016.pdf.

3 Author's calculations. The UK beer market is estimated at $15.5 billion in 2019, per https://www.statista.com/outlook/10010000/156/beer/ united-kingdom; meat products, $18.6 billion in 2019, per https:// www.statista.com/outlook/40020000/156/meat-products-sausages/ united-kingdom; bread, $4.7 billion in 2019, per https://www.statista. com/outlook/40050100/156/bread/united-kingdom. UK GDP is estimated at $2,800 billion in 2018, per https://en.wikipedia.org/wiki/ Economy_of_the_United_Kingdom.

4 https://en.wikipedia.org/wiki/Tithe.

5 Adrian Sargeant and Elaine Jay, *Fundraising Management: Analysis, Planning and Practice* (Abingdon: Routledge, 2014).

6 Scott M. Cutlip, *Fund Raising in the United States: Its Role in America's Philanthropy* (Piscataway, NJ: Transaction Publishers, 1965).

7 Cutlip, *Fund Raising.*

8 Sargeant and Jay, *Fundraising Management.*

9 Sargeant and Jay, *Fundraising Management.*

10 *The Rotarian*, October 1924, available at https://books.google.co.uk/ books?id=TUQEAAAAMBAJ&pg=PA59.

11 Anna Isaac, 'Have charity shock ads lost their power to disturb?', *Guardian*, 20 April 2016, https://www.theguardian.com/ voluntary-sector-network/2016/apr/20/charity-ads-shock-barnados.

12 See e.g. Amihai Glazer and Kai A. Konrad, 'A Signaling Explanation for Charity', *American Economic Review* 86.4 (September 1996): 1019–28, https:// www.jstor.org/stable/2118317.

13 Geoffrey Miller, *The Mating Mind: How Sexual Choice Shaped the Evolution of Human Nature* (London: Vintage, 2000).

14 Craig Landry, Andreas Lange, John A. List, Michael K. Price and Nicholas G. Rupp, 'Toward an Understanding of the Economics of Charity: Evidence from a Field Experiment', East Carolina University, University of Chicago, University of Maryland, University of Nevada-Reno, NBER and RFF, 2005, http://www.chicagocdr.org/papers/listpaper. pdf#search=%2522towards%20an%20understanding%20of%20the%20 economics%20of%20charity%2522.

15 James Andreoni, 'Impure Altruism and Donations to Public Goods: A Theory of Warm Glow Giving', *Economic Journal* 100.401 (June 1990): 464–77, available at https://econweb.ucsd.edu/~jandreon/Publications/ ej90.pdf.

16 'Introduction to Effective Altruism', Centre for Effective Altruism, 22 June

2016, https://www.effectivealtruism.org/articles/introduction-to-effective-altruism/.

17 https://givewell.org.

18 Dean Karlan and Daniel Wood, 'The Effect of Effectiveness: Donor Response to Aid Effectiveness in a Direct Mail Fundraising Experiment', *Economic Growth Center Discussion Paper No. 1038*, 2015.

19 'Mega-charities', The GiveWell Blog, 28 December 2011, https://blog.givewell.org/2011/12/28/mega-charities/.

16 Santa Claus

1 Lindsay Whipp, 'All Japan Wants for Christmas Is Kentucky Fried Chicken', *Financial Times*, 19 December 2010, https://www.ft.com/content/bb2dafc6-0ba4-11e0-a313-00144feabdc0#axzz2F2h70NMo.

2 https://www.theguardian.com/lifeandstyle/2016/dec/21/coca-cola-didnt-invent-santa-the-10-biggest-christmas-myths-debunked; https://www.coca-colacompany.com/stories/coke-lore-santa-claus.

3 http://www.whiterocking.org/santa.html.

4 https://www.snopes.com/fact-check/rudolph-red-nosed-reindeer/.

5 Stephen Nissenbaum, *The Battle for Christmas* (New York: Alfred Knopf, 1997); Bruce David Forbes, *America's Favorite Holidays: Candid Histories* (Oakland: University of California Press, 2015).

6 Nissenbaum, *Battle for Christmas*.

7 Forbes, *America's Favorite Holidays*.

8 John Tierney, 'The Big City, Christmas, and the Spirit of Commerce', *New York Times*, 21 December 2001, https://www.nytimes.com/2001/12/21/nyregion/the-big-city-christmas-and-the-spirit-of-commerce.html.

9 Joel Waldfogel, *Scroogenomics: Why You Shouldn't Buy Presents for the Holidays* (Woodstock: Princeton University Press, 2009).

10 Waldfogel, *Scroogenomics*.

11 Harriet Beecher Stowe, 'Christmas; or, The Good Fairy', *National Era* 4 (26 December 1850).

12 Marla Frazee, *Santa Claus: The World's Number One Toy Expert* (Boston, MA: Houghton Mifflin Harcourt, 2005).

13 Joel Waldfogel, 'The Deadweight Loss of Christmas', *American Economic Review* 83.5 (1993): 1328–36, www.jstor.org/stable/2117564.

14 Parag Waknis and Ajit Gaikwad, 'The Deadweight Loss of Diwali', *MPRA Paper*, University Library of Munich, 2011, https://EconPapers.repec.org/RePEc:pra:mprapa:52883.

15 The World Bank includes the IBRD (which makes non-concessional loans) and IDA (which makes concessional loans), and each lent about $20 billion in 2017. http://pubdocs.worldbank.org/en/982201506096253267/AR17-World-Bank-Lending.pdf.

16 Jennifer Pate Offenberg, 'Markets: Gift Cards', *Journal of Economic Perspectives* 21.2 (Spring 2007).

17 Francesca Gino and Francis J. Flynn, 'Give them what they want:

The benefits of explicitness in gift exchange', *Journal of Experimental Social Psychology* 47 (2011): 915–22, https://static1.squarespace.com/static/55dcde36e4b0df55a96ab220/t/55e746dee4b07156fbd7f6bd/1441220318875/Gino+Flynn+JESP+2011.pdf.

17 SWIFT

1   Susan Scott and Markos Zachariadis, *The Society for Worldwide Interbank Financial Telecommunication (SWIFT)* (Abingdon: Routledge, 2014), p. 12.
2   Tom Standage, *The Victorian Internet* (London: Weidenfeld and Nicolson, 1998), pp. 110–11.
3   Patrice A. Carré, 'From the telegraph to the telex: a history of technology, early networks and issues in France in the 19th and 20th centuries', *FLUX Cahiers scientifiques internationaux Réseaux et Territoires* 11 (1993): 17–31.
4   Eric Sepkes, quoted in Scott and Zachariadis, *The Society*, pp. 11–12.
5   The words of the Italian banker Renato Polo, quoted in Scott and Zachariadis, *The Society*, p. 18.
6   Scott and Zachariadis, *The Society*, p. 19.
7   'New SWIFT network gives banks an instant linkup – worldwide', *Banking* 69.7 (1977): 48.
8   Scott and Zachariadis, *The Society*, chs. 2–3.
9   Lily Hay Newman, 'A New Breed of ATM Hackers Gets in Through a Bank's Network', *Wired*, 9 April 2019; Iain Thomson, 'Banking system SWIFT was anything but on security, ex-boss claims', The Register, 18 August 2016, https://www.theregister.co.uk/2016/08/18/swift_was_anything_but_on_security_claim/.
10   Michael Peel and Jim Brunsden, 'Swift shows impact of Iran dispute on international business', *Financial Times*, 6 June 2018, https://www.ft.com/content/9f082a96-63f4-11e8-90c2-9563a0613e56.
11   Eric Lichtblau and James Risen, 'Bank Data Is Sifted by U.S. in Secret to Block Terror', *New York Times*, 23 June 2006.
12   Michael Peel, 'Swift to comply with US sanctions on Iran in blow to EU', *Financial Times*, 5 November 2018, https://www.ft.com/content/8f16f8aa-e104-11e8-8e70-5e22a430c1ad.
13   Justin Scheck and Bradley Hope, 'The Dollar Underpins American Power', *Wall Street Journal*, 29 May 2019.
14   Henry Farrell and Abraham L. Newman, 'Weaponized Interdependence: How Global Economic Networks Shape State Coercion', *International Security* 2019 44:1, 42–79.
15   Nicholas Lambert, *Planning Armageddon* (London: Harvard University Press, 2012).

18 Credit Cards

1   Lewis Mandell, *The Credit Card Industry: A History* (Boston, MA: Twayne Publishers, 1990), p. xii; The Department Store Museum website, http://

www.thedepartmentstoremuseum.org/2010/11/charge-cards.html; Hilary Greenbaum and Dana Rubinstein, 'The Cardboard Beginnings of the Credit Card', *New York Times*, 2 December 2011, http://www.nytimes.com/2011/12/04/magazine/the-cardboard-beginnings-of-the-credit-card.html.

2   Mandell, *Credit Card Industry*, p. 26; Greenbaum and Rubinstein, 'Cardboard Beginnings'.

3   David S. Evans and Richard Schmalensee, *Paying with Plastic: The Digital Revolution in Buying and Borrowing* (Cambridge, MA: MIT Press, 1999), p. 79.

4   Bank of America, 'Our History', https://about.bankofamerica.com/en-us/our-story/birth-of-modern-credit-card.html; *99 Percent Invisible*, 'The Fresno Drop', Episode 196, https://99percentinvisible.org/episode/the-fresno-drop/.

5   History of IBM, http://www-03.ibm.com/ibm/history/ibm100/us/en/icons/magnetic/.

6   Evans and Schmalensee, *Paying with Plastic*, pp. 7–9.

7   Maddy Savage, 'Why Sweden is close to becoming a cashless economy', BBC News, 12 September 2017, http://www.bbc.co.uk/news/business-41095004.

8   Mandell, *Credit Card Industry*, p. 39.

9   Drazen Prelec and Duncan Simester, 'Always Leave Home Without It: A Further Investigation of the Credit-Card Effect on Willingness to Pay', *Marketing Letters* 12.1 (2001): 5–12, http://web.mit.edu/simester/Public/Papers/Alwaysleavehome.pdf.

10   Thomas A. Durkin, *Consumer Credit and the American Economy* (Oxford: Oxford University Press, 2014), p. 267.

11   https://blogs.imf.org/2017/10/03/rising-household-debt-what-it-means-for-growth-and-stability/.

12   Durkin, *Consumer Credit*, Table 7.7, pp. 312–23.

19 Stock Options

1   https://www.c-span.org/video/?23518-1/clinton-campaign-speech (starts at 27:14).

2   Planet Money, 'Episode 682: When CEO Pay Exploded', 5 February 2016, https://www.npr.org/templates/transcript/transcript.php?storyId=465747726.

3   Lawrence Mishel and Jessica Schieder, 'CEO compensation surged in 2017', Economic Policy Institute, Washington, DC, 16 August 2018, epi.org/152123.

4   Aristotle, *Politics* 1.11.

5   An alternative possibility is that Thales may have entered into a legally binding contract to hire the presses: if so, he would have invented not the option but the future. See George Crawford and Bidyut Sen, *Derivatives for Decision Makers: Strategic Management Issues* (Hoboken, NJ: John Wiley & Sons, 1996), p. 7.

6 Aristotle, *Politics* 1.11.

7 Crawford and Sen, *Derivatives*, p. 7.

8 Crawford and Sen, *Derivatives*, p. 20.

9 Michael C. Jensen and Kevin J. Murphy, 'CEO Incentives – It's Not How Much You Pay, But How', *Harvard Business Review* 3 (May–June 1990): 138–53.

10 Robert Reich, 'There's One Big Unfinished Promise By Bill Clinton that Hillary Should Put to Bed', 7 September 2016, https://robertreich.org/post/150082237740.

11 Brian J. Hall and Kevin J. Murphy, 'The Trouble with Stock Options', NBER Working Paper No. 9784, June 2003.

12 Jerry W. Markham and Rigers Gjyshi (eds), *Research Handbook on Securities Regulation in the United States* (Cheltenham and Northampton, MA: Edward Elgar Publishing, 2014), p. 254.

13 Hall and Murphy, 'The Trouble'.

14 Lucian A. Bebchuk and Jesse M. Fried, 'Pay without Performance: The Unfulfilled Promise of Executive Compensation', Harvard Law School John M. Olin Center for Law, Economics and Business Discussion Paper Series, Paper 528, 2003, p. 10.

15 Bebchuk and Fried, 'Pay without Performance', p. 67.

16 Bebchuk and Fried, 'Pay without Performance'. See also, more recently, Indira Tulepova, 'The Impact of Ownership Structure on CEO Compensation: Evidence from the UK', MA thesis, Radboud University Nijmegen Faculty of Management, 2016–17.

17 Bebchuk and Fried, 'Pay without Performance'. See also Tulepova, 'Impact of Ownership Structure'.

18 Marianne Bertrand and Sendhil Mullainathan, 'Are CEOs Rewarded for Luck? The Ones Without Principals Are', *Quarterly Journal of Economics* 116.3 (August 2001): 901–32, https://doi.org/10.1162/00335530152466269.

19 Rosanna Landis Weaver, 'The Most Overpaid CEOs: Are Fund Managers Asleep at The Wheel?', Harvard Law School Forum on Corporate Governance and Financial Regulation, 30 March 2019, https://corpgov.law.harvard.edu/2019/03/30/the-most-overpaid-ceos-are-fund-managers-asleep-at-the-wheel/.

20 Fernando Duarte, 'It takes a CEO just days to earn your annual wage', BBC, 9 January 2019, http://www.bbc.com/capital/story/20190108-how-long-it-takes-a-ceo-to-earn-more-than-you-do-in-a-year.

21 See e.g. Mishel and Schieder, 'CEO compensation surged in 2017'.

22 See e.g. Robert C. Pozen and S. P. Kothari, 'Decoding CEO Pay', *Harvard Business Review*, July–August 2017, https://hbr.org/2017/07/decoding-ceo-pay; Nicholas E. Donatiello, David F. Larcker and Brian Tayan, 'CEO Talent: A Dime a Dozen, or Worth its Weight in Gold?', Stanford Closer Look Series, September 2017, https://www.gsb.stanford.edu/faculty-research/publications/ceo-talent-dime-dozen-or-worth-its-weight-gold.

23 Mishel and Schieder, 'CEO compensation surged in 2017'.

24   See e.g. David F. Larcker, Nicholas E. Donatiello and Brian Tayan, 'Americans and CEO Pay: 2016 Public Perception Survey on CEO Compensation', Corporate Governance Research Initiative, Stanford Rock Center for Corporate Governance, February 2016, https://www.gsb. stanford.edu/faculty-research/publications/americans-ceo-pay-2016-public-perception-survey-ceo-compensation; Dina Gerdeman, 'If the CEO's High Salary Isn't Justified to Employees, Firm Performance May Suffer', Working Knowledge, Harvard Business School, 17 January 2018, https://hbswk.hbs. edu/item/if-the-ceo-s-high-salary-isn-t-justified-to-employees-firm-performance-may-suffer.

20 The Vickrey Turnstile

1   Jacques H. Drèze, *William S. Vickrey 1914–1996: A Biographical Memoir* (Washington, DC: National Academies Press, 1998), http://www.nasonline. org/publications/biographical-memoirs/memoir-pdfs/vickrey-william.pdf; Ronald Harstad, 'William S. Vickrey', https://economics.missouri.edu/ working-papers/2005/wp0519_harstad.pdf.

2   Yohana Desta, '1904 to today: See how New York City subway fare has climbed over 111 years', Mashable, 22 March 2015, https://mashable. com/2015/03/22/new-york-city-subway-fare/.

3   William S. Vickrey, 'The revision of the rapid transit fare structure of the City of New York: Finance project', New York, 1952.

4   Vickrey, 'The revision'.

5   Daniel Levy and Andrew T. Young, '"The Real Thing": Nominal Price Rigidity of the Nickel Coke, 1886–1959', *Journal of Money, Credit and Banking* 36.4 (August 2004): 765–99, available at SSRN: https://ssrn.com/ abstract=533363.

6   William S. Vickrey, 'My Innovative Failures in Economics', *Atlantic Economic Journal* 21 (1993): 1–9.

7   Jaya Saxena, 'The Extinction of the Early Bird', Eater, 29 January 2018, https://www.eater.com/2018/1/29/16929816/early-bird-extinction-florida.

8   John Koten, 'Fare Game: In Airlines' Rate War, Small Daily Skirmishes Often Decide Winners', *Wall Street Journal*, 24 August 1984.

9   Koten, 'Fare Game'.

10   Dug Begley, 'Almost $250 for 13 miles: Uber's "surge pricing"', *Houston Chronicle*, 30 December 2014.

11   Nicholas Diakopoulos, 'How Uber surge pricing really works', *Washington Post*, 17 April 2015.

12   Daniel Kahneman, Jack L. Knetsch and Richard Thaler, 'Fairness as a Constraint on Profit Seeking: Entitlements in the Market', *American Economic Review* 76.4 (1986): 728–41, http://www.jstor.org/ stable/1806070.

13   Constance L. Hays, 'Variable-Price Coke Machine Being Tested', *New York Times*, 28 October 1999; David Leonhardt, 'Airline Tickets Can Be More in June Than in January. But Soda? Forget It', *New York Times*, 27 June 2005.

14 Robin Harding, 'Rail privatisation: the UK looks for secrets of Japan's success', *Financial Times*, 28 January 2019, https://www.ft.com/content/9f7f044e-1f16-11e9-b2f7-97e4dbd3580d.

15 David Schaper, 'Are $40 road tolls the future?', *NPR All Things Considered*, 12 December 2017, https://www.npr.org/2017/12/12/570248568/are-40-toll-roads-the-future.

16 Harstad, 'William S. Vickrey'.

21 The Blockchain

1 Arie Shapira and Kailey Leinz, 'Long Island Iced Tea Soars After Changing Its Name to Long Blockchain', *Bloomberg*, 21 December 2017, https://www.bloomberg.com/news/articles/2017-12-21/crypto-craze-sees-long-island-iced-tea-rename-as-long-blockchain.

2 Jason Rowley, 'With at least $1.3 billion invested globally in 2018, VC funding for blockchain blows past 2017 totals', TechCrunch, 20 May 2018, https://techcrunch.com/2018/05/20/with-at-least-1-3-billion-invested-globally-in-2018-vc-funding-for-blockchain-blows-past-2017-totals.

3 Jonathan Chester, 'What You Need To Know About Initial Coin Offering Regulations', *Forbes*, 9 April 2018, https://www.forbes.com/sites/jonathanchester/2018/04/09/what-you-need-to-know-about-initial-coin-offering-regulations.

4 https://bitcoin.org/bitcoin.pdf.

5 See e.g. 'How blockchains could change the world', interview with Don Tapscott, *McKinsey*, May 2016, https://www.mckinsey.com/industries/high-tech/our-insights/how-blockchains-could-change-the-world; Laura Shin, 'How The Blockchain Will Transform Everything From Banking To Government To Our Identities', *Forbes*, 26 May 2016, https://www.forbes.com/sites/laurashin/2016/05/26/how-the-blockchain-will-transform-everything-from-banking-to-government-to-our-identities/.

6 Christian Catalini and Joshua Gans, 'Some Simple Economics of the Blockchain', Rotman School of Management Working Paper No. 2874598, MIT Sloan Research Paper No. 5191-16, https://papers.ssrn.com/sol3/papers.cfm?abstract_id=2874598.

7 Steven Johnson, 'Beyond the Bitcoin Bubble', *New York Times*, 16 January 2018, https://www.nytimes.com/2018/01/16/magazine/beyond-the-bitcoin-bubble.html.

8 John Biggs, 'Exit scammers run off with $660 million in ICO earnings', TechCrunch, 13 April 2018, https://techcrunch.com/2018/04/13/exit-scammers-run-off-with-660-million-in-ico-earnings/.

9 Tyler Cowen, 'Don't Let Doubts About Blockchains Close Your Mind', *Bloomberg*, 27 April 2018, https://www.bloomberg.com/view/articles/2018-04-27/blockchains-warrant-skepticism-but-keep-an-open-mind.

10 Jan Vermeulen, 'Bitcoin and Ethereum vs Visa and PayPal – Transactions per second', My Broadband, 22 April 2017, https://mybroadband.co.za/

news/banking/206742-bitcoin-and-ethereum-vs-visa-and-paypal-
transactions-per-second.html.

11   Alex de Vries, 'Bitcoin's Growing Energy Problem', *Joule* 2.5 (16 May 2018):
     801–5, https://www.cell.com/joule/fulltext/S2542-4351(18)30177-6.

12   Preethi Kasireddy, 'Blockchains don't scale. Not today, at least. But there's
     hope', Hacker Noon, 22 August 2017, https://hackernoon.com/
     blockchains-dont-scale-not-today-at-least-but-there-s-hope-2cb43946551a.

13   Catherine Tucker and Christian Catalini, 'What Blockchain Can't
     Do', *Harvard Business Review*, 28 June 2018, https://hbr.org/2018/06/
     what-blockchain-cant-do.

14   Kai Stinchcombe, 'Ten years in, nobody has come up with a use for
     blockchain', Hacker Noon, 22 December 2017, https://hackernoon.com/
     ten-years-in-nobody-has-come-up-with-a-use-case-for-blockchain-ee
     98c180100.

15   Mark Frauenfelder, '"I Forgot My PIN": An Epic Tale of Losing $30,000 in
     Bitcoin', *Wired*, 10 October 2017, https://www.wired.com/story/i-forgot-
     my-pin-an-epic-tale-of-losing-dollar30000-in-bitcoin/.

16   Chris Wray, 'Law and global trade in the era of blockchain', 9 April 2018,
     https://medium.com/humanizing-the-singularity/law-and-global-trade-in-
     the-era-of-blockchain-2695c6276579.

17   Klint Finley, 'A $50 Million Hack Just Showed That the DAO Was All Too
     Human', *Wired*, 18 June 2016, https://www.wired.com/2016/06/50-
     million-hack-just-showed-dao-human/.

18   Eric Budish, 'The Economic Limits of Bitcoin and the Blockchain',
     5 June 2018, http://faculty.chicagobooth.edu/eric.budish/research/
     Economic-Limits-Blockchain.pdf.

19   Jim Edwards, 'One of the kings of the '90s dot-com bubble now faces
     20 years in prison', Business Insider, 6 December 2016,
     http://uk.businessinsider.com/where-are-the-kings-of-the-1990s-
     dot-com-bubble-bust-2016-12/.

20   Josiah Wilmoth, 'Ex-Iced Tea Maker "Long Blockchain" Faces Reckoning
     as Nasdaq Prepares to Delist Its Shares', CCN, 6 June 2018, https://www.
     ccn.com/ex-iced-tea-maker-long-blockchain-faces-reckoning-as-nasdaq-
     prepares-to-delist-its-shares/.

22 Interchangeable Parts

1   Simon Winchester, *Exactly: How Precision Engineers Created the Modern World*
    (London: William Collins, 2018), pp. 90–4.

2   Marshall Brain, 'How Flintlock Guns Work', https://science.howstuffworks.
    com/flintlock2.htm.

3   Winchester, *Exactly*, pp. 90–4.

4   William Howard Adams, *The Paris Years of Thomas Jefferson* (New Haven,
    CT: Yale University Press, 1997).

5   Thomas Jefferson, 30 August 1785, in *The Papers of Thomas Jefferson*, ed.
    Julian Boyd (Princeton, NJ: Princeton, University Press, 1950).

6   David A. Hounshell, *From the American System to Mass Production, 1800–1932* (Baltimore, MD: Johns Hopkins Press, 1984), p. 26.
7   Frank Dawson, *John Wilkinson: King of the Ironmasters* (Cheltenham: The History Press, 2012).
8   Winchester, *Exactly*.
9   H. W. Dickinson, *A Short History of the Steam Engine* (Cambridge: Cambridge University Press, 1939), ch. V.
10 Robert C. Allen, *Global Economic History: A Very Short Introduction* (Oxford: Oxford University Press, 2011), ch. 3.
11 L. T. C. Rolt, *Tools for the Job* (London: HM Stationery Office, 1986), pp. 55–63; Ben Russell, *James Watt* (London: Reaktion Books, 2014), pp. 129–30.
12 Winchester, *Exactly*.
13 Adam Smith, *An Inquiry into the Nature and Causes of the Wealth of Nations* (1776), book 1, p. 1, available at https://www.econlib.org/library/Smith/smWN.html?chapter_num=4#book-reader.
14 Hounshell, *American System to Mass Production*, p. 3.
15 Priya Satia, *Empire of Guns* (London: Duckworth Overlook, 2018), pp. 353–5.
16 Hounshell, *American System to Mass Production*; Winchester, *Exactly*.
17 Hounshell, *American System to Mass Production*; Winchester, *Exactly*.

23 RFID

1   Adam Fabio, 'Theremin's Bug: How the Soviet Union Spied on the US Embassy for Seven Years', Hackaday.com, 8 December 2015, https://hackaday.com/2015/12/08/theremins-bug/.
2   Martin Vennard, 'Leon Theremin: The man and the music machine', BBC World Service, 13 March 2012, https://www.bbc.co.uk/news/magazine-17340257.
3   A.W., 'RFIDs are set almost to eliminate lost luggage', *Economist*, 1 November 2016, https://www.economist.com/gulliver/2016/11/01/rfids-are-set-almost-to-eliminate-lost-luggage.
4   Bill Glover and Himanshu Bhatt, *RFID Essentials* (Sebastopol, CA: O'Reilly, 2006).
5   Jordan Frith, *A Billion Little Pieces: RFID and Infrastructures of Identification* (Cambridge, MA: MIT Press, 2019).
6   'Radio silence', *Economist Technology Quarterly*, 7 June 2007, https://www.economist.com/technology-quarterly/2007/06/07/radio-silence.
7   Glover and Bhatt, *RFID Essentials*; Frith, *A Billion Little Pieces*.
8   Jonathan Margolis, 'I am microchipped and have no regrets', *Financial Times*, 31 May 2018, https://www.ft.com/content/6c0591b4-632d-11e8-bdd1-cc0534df682c.
9   Kevin Ashton, 'That "Internet of Things" Thing', *RFID Journal*, 22 June 2009, https://www.rfidjournal.com/articles/view?4986.
10 N.V., 'The Difference Engine: Chattering Objects', *Economist*, 13 August 2010, https://www.economist.com/babbage/2010/08/13/the-difference-engine-chattering-objects.

11   Cory Doctorow, 'Discarded smart lightbulbs reveal your wifi passwords, stored in the clear', *BoingBoing*, 29 January 2019, https://boingboing. net/2019/01/29/fiat-lux.html.

12   https://www.pentestpartners.com/security-blog/gps-watch-issues-again/.

13   'Privacy Not Included: Vibratissimo Panty Buster', Mozilla Foundation, 1 November 2018, https://foundation.mozilla.org/en/privacynotincluded/ products/vibratissimo-panty-buster/.

14   Shoshana Zuboff, *The Age of Surveillance Capitalism* (London: Profile, 2019); Bruce Sterling, *The Epic Struggle of the Internet of Things* (Moscow: Strelka Press, 2014).

24 The Interface Message Processor

1   Katie Hafner and Matthew Lyon, *Where Wizards Stay Up Late* (New York: Touchstone, 1996), p. 22.

2   Hafner and Lyon, *Wizards*, pp. 10–15.

3   Hafner and Lyon, *Wizards*, pp. 10–15.

4   Hafner and Lyon, *Wizards*, p. 42.

5   Gene I. Rochlin, *Trapped in the Net* (Princeton, NJ: Princeton University Press, 1997), pp. 38–40. Also, personal communication with Adrian Harford, a veteran computer engineer.

6   Janet Abbate, *Inventing the Internet* (Cambridge, MA: MIT Press, 1999), p. 48.

7   Peter H. Salus, *Casting the Net: From ARPANET to Internet and Beyond* (Reading, MA: Addison-Wesley, 1995), p. 21. Clark himself modestly commented that 'someone else would have thought of it in a few days or weeks'.

8   Graham Linehan, *The IT Crowd*: 'The Speech' (aired December 2008; https://www.imdb.com/title/tt1320786/).

9   Abbate, *Inventing*, pp. 52–3; for a description of how more modern routers work, see Andrew Blum, *Tubes* (London: Viking, 2012), pp. 29–30.

10   Old Computers, http://www.old-computers.com/museum/computer. asp?c=551.

11   Blum, *Tubes*, p. 39. (David Bunnell, *Making the Cisco Connection* (New York: John Wiley, 2000), p. 4, puts the cost at $100,000.)

12   Abbate, *Inventing*, pp. 62–3.

13   http://www.historyofinformation.com/expanded.php?id=1108; Hafner and Lyon, *Wizards*, pp. 150–3.

14   Abbate, *Inventing*, pp. 194–5.

25 GPS

1   'Swedes miss Capri after GPS gaffe', BBC, 28 July 2009, http://news.bbc. co.uk/1/hi/world/europe/8173308.stm.

2   'Economic impact to the UK of a disruption to GNSS', Showcase Report, April 2017, https://assets.publishing.service.gov.uk/government/uploads/

system/uploads/attachment_data/file/619545/17.3254_Economic_impact_ to_UK_of_a_disruption_to_GNSS_-_Showcase_Report.pdf.

3 Greg Milner, *Pinpoint: How GPS Is Changing Our World* (London: Granta, 2016).

4 Dan Glass, 'What Happens If GPS Fails?', *Atlantic*, 13 June 2016, https:// www.theatlantic.com/technology/archive/2016/06/what-happens-if-gps- fails/486824/.

5 William Jackson, 'Critical infrastructure not prepared for GPS disruption', GCN, 8 November 2013, https://gcn.com/articles/2013/11/08/ gps-disruption.aspx.

6 Milner, *Pinpoint*.

7 Milner, *Pinpoint*.

8 Milner, *Pinpoint*.

9 Milner, *Pinpoint*.

10 Milner, *Pinpoint*.

11 Victoria Woollaston, 'Solar storms 2018: What is a solar storm and when will the next one hit Earth?', Alphr, 12 April 2018, http://www.alphr.com/ science/1008518/solar-storm-earth-charged-particles.

12 Milner, *Pinpoint*.

13 Milner, *Pinpoint*.

14 'National Risk Estimate', Department of Homeland Security, available at https://rntfnd.org/wp-content/uploads/DHS-National-Risk-Estimate-GP S-Disruptions.pdf.

15 Milner, *Pinpoint*.

16 Greg Milner, 'What Would Happen If G.P.S. Failed?', *New Yorker*, 6 May 2016, https://www.newyorker.com/tech/elements/what-would-happen-if-gps-failed.

26 The Movable-Type Printing Press

1 Frédéric Barbier, *Gutenberg's Europe: The Book and the Invention of Western Modernity* (London: Polity Press, 2016).

2 John Man, *The Gutenberg Revolution* (London: Bantam, 2009), ch. 2.

3 Julie Mellby, 'One Million Buddhist Incantations', 3 January 2009, https:// www.princeton.edu/~graphicarts/2009/01/one_million_buddhist_ incantati.html.

4 Tom Scocca, 'The first printed books came with a question: What do you do with these things?', *Boston Globe*, 29 August 2010, http://www.boston. com/bostonglobe/ideas/articles/2010/08/29/cover_story/?page=full.

5 Mary Wellesley, 'Gutenberg's printed Bible is a landmark in European culture', *Apollo Magazine* 8 (September 2018), https://www.apollo-magazine. com/gutenbergs-printed-bible-landmark-european-culture/; 'Fifty Treasures: The Gutenberg Bible', https://www.50treasures.divinity.cam. ac.uk/treasure/gutenberg-bible/.

6 Jeremiah Dittmar, 'Europe's Transformation After Gutenberg', *Centrepiece* 544 (Spring 2019).

7 John Naughton, *From Gutenberg to Zuckerberg: What You Really Need to Know About the Internet* (London: Quercus, 2012), ch. 1.

8   Dittmar, 'Europe's Transformation'.
9   Simon Winchester, *Exactly*.
10 Dittmar, 'Europe's Transformation'.
11 Barbier, *Gutenberg's Europe*.
12 Paul Ormerod, *Why Most Things Fail* (London: Faber & Faber, 2005), p. 15.
13 Elizabeth Eisenstein, *The Printing Revolution in Early Modern Europe* (New York: Cambridge University Press, 1983).
14 Eisenstein, *Printing Revolution*.
15 Andrew Marantz, *Antisocial: Online Extremists, Techno-Utopians, and the Hijacking of the American Conversation* (New York: Viking, 2019).
16 https://www.bl.uk/treasures/gutenberg/basics.html.
17 Scocca, 'The first printed books', *Boston Globe*, 29 August 2010.

27 Menstrual Pads

1   Sharra Vostral, *Under Wraps* (Plymouth: Lexington Books, 2011), ch. 4.
2   Vostral, *Under Wraps*, ch. 1.
3   Thomas Heinrich and Bob Batchelor, *Kotex, Kleenex, Huggies: Kimberly-Clark and the Consumer Revolution in American Business* (Columbus: Ohio State University Press, 2004), p. 96.
4   Janice Delaney, Mary Jane Lupton, Emily Toth, *The Curse: A Cultural History of Menstruation* (New York: New American Library, 1976).
5   Vostral, *Under Wraps*, ch. 3.
6   Delaney et al., *The Curse*; Ashley Fetters, 'The Tampon: A History', *Atlantic*, June 2015, https://www.theatlantic.com/health/archive/2015/06/history-of-the-tampon/394334/.
7   A. Juneja, A. Sehgal, A. B. Mitra, A. Pandey, 'A Survey on Risk Factors Associated with Cervical Cancer', *Indian Journal of Cancer* 40.1 (January–March 2003): 15–22, https://www.ncbi.nlm.nih.gov/pubmed/14716127; Colin Schultz, 'How Taboos Around Menstruation Are Hurting Women's Health', *Smithsonian Magazine*, 6 March 2014, https://www.smithsonianmag.com/smart-news/how-taboos-around-menstruation-are-hurting-womens-health-180949992/.
8   Delaney et al., *The Curse*; Museum of Menstruation website, http://www.mum.org/collection.htm.
9   Kat Eschner, 'The Surprising Origins of Kotex Pads', *Smithsonian Magazine*, 11 August 2017, https://www.smithsonianmag.com/innovation/surprising-origins-kotex-pads-180964466/.
10 Eschner, 'Surprising Origins'.
11 Eschner, 'Surprising Origins'.
12 Vostral, *Under Wraps*, ch. 4.
13 Fetters, 'The Tampon'.
14 Kelly O'Donnell, 'The whole idea might seem a little strange to you: Selling the menstrual cup', *Technology Stories*, 4 December 2017, https://www.technologystories.org/menstrual-cups/.
15 Delaney et al., *The Curse*.

16 Susan Dudley, Salwa Nassar, Emily Hartman and Sandy Wang, 'Tampon Safety', National Center for Health Research, http://www.center4research.org/tampon-safety/. They cite a 2015 Euromonitor report.

17 Andrew Adam Newman, 'Rebelling Against the Commonly Evasive Feminine Care Ad', *New York Times*, 16 March 2010.

18 Vibeke Venema, 'The Indian sanitary pad revolutionary', *BBC Magazine*, 4 March 2014.

19 Oni Lusk-Stover, Rosemary Rop, Elaine Tinsely and Tamer Samah Rabie, 'Globally, periods are causing girls to be absent from school', World Bank Blog: Education for Global Development, 27 June 2016, https://blogs.worldbank.org/education/globally-periods-are-causing-girls-be-absent-school.

20 Thomas Friedman, 'Cellphones, Maxi-Pads and Other Life-Changing Tools', *New York Times*, 6 April 2007.

28 CCTV

1 Albert Abramson, *The History of Television, 1942 to 2000* (Jefferson: McFarland, 2002).

2 Michael Marek, 'The V-2: the first space rocket', Deutsche Welle, 2 October 2012, https://www.dw.com/en/the-v-2-the-first-space-rocket/a-16276064.

3 Bob Ward, *Dr. Space: The Life of Wernher Von Braun* (Annapolis: Naval Institute Press, 2005).

4 Abramson, *History of Television*.

5 Niall Jenkins, '245 million video surveillance cameras installed globally in 2014', IHS Markit, 11 June 2015, https://technology.ihs.com/532501/245-million-video-surveillance-cameras-installed-globally-in-2014.

6 Frank Hersey, 'China to have 626 million surveillance cameras within 3 years', Technode, 22 November 2017, https://technode.com/2017/11/22/china-to-have-626-million-surveillance-cameras-within-3-years/.

7 Dan Strumpf, Natasha Khan and Charles Rollet, 'Surveillance Cameras Made by China Are Hanging All Over the U.S.', *Wall Street Journal*, 12 November 2017, https://www.wsj.com/articles/surveillance-cameras-made-by-china-are-hanging-all-over-the-u-s-1510513949.

8 Paul Mozur, 'Inside China's Dystopian Dreams: A.I., Shame and Lots of Cameras', *New York Times*, 8 July 2018, https://www.nytimes.com/2018/07/08/business/china-surveillance-technology.html.

9 Matthew Carney, 'Leave no dark corner', ABC, 17 September 2018, http://www.abc.net.au/news/2018-09-18/china-social-credit-a-model-citizen-in-a-digital-dictatorship/10200278.

10 Simina Mistreanu, 'Life Inside China's Social Credit Laboratory', *Foreign Policy*, 3 April 2018, https://foreignpolicy.com/2018/04/03/life-inside-chinas-social-credit-laboratory/.

11 Mistreanu, 'China's Social Credit Laboratory'.

12 Mistreanu, 'China's Social Credit Laboratory'.

13 Mistreanu, 'China's Social Credit Laboratory'.

14 Henry Cowles, 'Orwell knew: we willingly buy the screens that are used against us', Aeon, 24 July 2018, https://aeon.co/ideas/orwell-knew-we-willingly-buy-the-screens-that-are-used-against-us.

15 https://www.smbc-comics.com/comic/listening.

16 Scott Carey, 'Does Amazon Alexa or Google Home listen to my conversations?', TechWorld, 25 May 2018, https://www.techworld.com/security/does-amazon-alexa-listen-to-my-conversations-3661967/.

17 Carney, 'Leave no dark corner'.

18 https://en.wikipedia.org/wiki/Panopticon.

29 Pornography

1 Lyrics by Robert Lopez and Jeff Marx; book by Jeff Whitty.

2 Mark Ward, 'Web porn: Just how much is there?', BBC News, 1 July 2013, https://www.bbc.co.uk/news/technology-23030090.

3 https://www.alexa.com/topsites (accessed 24 September 2018). Netflix ranked 26th, Pornhub 28th, LinkedIn 29th.

4 R. Dale Guthrie, *The Nature of Paleolithic Art* (Chicago: University of Chicago Press, 2005).

5 http://www.britishmuseum.org/explore/a_history_of_the_world/objects.aspx#7.

6 Ilan Ben Zion, '4,000-year-old erotica depicts a strikingly racy ancient sexuality', *Times of Israel*, 17 January 2014, https://www.timesofisrael.com/4000-year-old-erotica-depicts-a-strikingly-racy-ancient-sexuality/.

7 April Holloway, 'Sex Pottery of Peru: Moche Ceramics Shed Light on Ancient Sexuality', 6 May 2015, https://www.ancient-origins.net/artifacts-other-artifacts/sex-pottery-peru-moche-ceramics-shed-light-ancient-sexuality-003017.

8 https://en.wikipedia.org/wiki/Kama_Sutra.

9 Patchen Barss, *The Erotic Engine: How Pornography Has Powered Mass Communication, from Gutenberg to Google* (Toronto: Doubleday Canada, 2010).

10 Barss, *Erotic Engine*.

11 https://www.etymonline.com/word/pornography.

12 Barss, *Erotic Engine*.

13 Eric Schlosser, *Reefer Madness: Sex, Drugs, and Cheap Labor in the American Black Market* (New York: HMH, 2004).

14 Barss, *Erotic Engine*.

15 Jonathan Coopersmith, 'Pornography, Videotape and the Internet', *IEEE Technology and Society Magazine*, Spring 2000.

16 Peter H. Lewis, 'Critics Troubled By Computer Study On Pornography', *New York Times*, 3 July 1995, https://www.nytimes.com/1995/07/03/business/critics-troubled-by-computer-study-on-pornography.html.

17 Barss, *Erotic Engine*.

18 Lewis Perdue, *EroticaBiz: How Sex Shaped the Internet* (Lincoln: Writers Club Press, 2002).

19 Joe Pinsker, 'The Hidden Economics of Porn', *Atlantic*, 4 April 2016, https://www.theatlantic.com/business/archive/2016/04/pornography-industry-economics-tarrant/476580/.

20 Jon Ronson, The Butterfly Effect, http://www.jonronson.com/butterfly.html.

21 Jon Ronson, 'Jon Ronson on bespoke porn: "Nothing is too weird. We consider all requests"', *Guardian*, 29 July 2017, https://www.theguardian.com/culture/2017/jul/29/jon-ronson-bespoke-porn-nothing-is-too-weird-all-requests.

22 David Auerbach, 'Vampire Porn', Slate, 23 October 2014, http://www.slate.com/articles/technology/technology/2014/10/mindgeek_porn_monopoly_its_dominance_is_a_cautionary_tale_for_other_industries.html.

23 https://www.youtube.com/watch?v=gTY1o0w_uEA.

24 Bruce Y. Lee, 'In Case You Are Wondering, Sex With Robots May Not Be Healthy', *Forbes*, 5 June 2018, https://www.forbes.com/sites/brucelee/2018/06/05/in-case-you-are-wondering-sex-with-robots-may-not-be-healthy.

30 Prohibition

1   Daniel Okrent, *Last Call: The Rise and Fall of Prohibition* (New York: Simon and Schuster, 2010).

2   Walter A. Friedman, *Fortune Tellers: The Story of America's First Economic Forecasters* (Princeton, NJ: Princeton University Press, 2013).

3   Mark Thornton, *The Economics of Prohibition* (Salt Lake City: University of Utah Press, 1991).

4   Okrent, *Last Call*.

5   Thornton, *Economics of Prohibition*.

6   Lisa McGirr, *The War on Alcohol: Prohibition and the Rise of the American State* (New York: WW Norton, 2015).

7   Mark Thornton, 'Cato Institute Policy Analysis No. 157: Alcohol Prohibition Was a Failure', Cato Institute, 1991, https://object.cato.org/sites/cato.org/files/pubs/pdf/pa157.pdf.

8   McGirr, *War on Alcohol*.

9   McGirr, *War on Alcohol*.

10 Thornton, *Economics of Prohibition*.

11 Thornton, *Economics of Prohibition*.

12 Gary S. Becker, 'Crime and Punishment: An Economic Approach', *Journal of Political Economy* 76.2 (1968): 169–217.

13 Tim Harford, 'It's the humanity, stupid: Gary Becker has lunch with the FT', 17 June 2006, http://timharford.com/2006/06/its-the-humanity-stupid-gary-becker-has-lunch-with-the-ft/.

14 Okrent, *Last Call*.

15 Okrent, *Last Call*.

16 Thornton, *Economics of Prohibition*.

17 Thornton, *Economics of Prohibition*.

18 Thornton, *Economics of Prohibition*.

19 Observation based on https://en.wikipedia.org/wiki/Prohibition (accessed 5 January 2019).

20 Eimor P. Santos, 'No alcohol, cockfights: What you can't do on May 14 election day', CNN Philippines, 12 May 2018, http://cnnphilippines.com/news/2018/05/12/Gun-ban-liquor-ban-what-you-cant-do-on-May-14-election-day.html.

21 'Alcohol sales ban tightened for Asanha Bucha, Lent', *The Nation*, 26 July 2018, http://www.nationmultimedia.com/detail/breakingnews/30350856.

22 Brian Wheeler, 'The slow death of prohibition', BBC News, 21 March 2012, https://www.bbc.co.uk/news/magazine-17291978.

23 https://en.wikipedia.org/wiki/Blue_laws_in_the_United_States.

24 'Revisiting Bootleggers and Baptists', Policy Report, Cato Institute, 17 September 2014, https://www.cato.org/policy-report/septemberoctober-2014/revisiting-bootleggers-baptists.

25 Philip Wallach and Jonathan Rauch, 'Bootleggers, Baptists, bureaucrats, and bongs: How special interests will shape marijuana legalization', Center for Effective Public Management at Brookings, June 2016, https://www.brookings.edu/wp-content/uploads/2016/07/bootleggers.pdf.

26 https://en.wikipedia.org/wiki/Legality_of_cannabis.

27 Christopher Snowdon, IEA Discussion Paper No. 90, 'Joint Venture: Estimating the Size and Potential of the UK Cannabis Market', Institute for Economic Affairs, 2018, https://iea.org.uk/wp-content/uploads/2018/06/DP90_Legalising-cannabis_web-1.pdf.

28 'Ending the Drug Wars', Report of the LSE Expert Group on the Economics of Drug Policy, London School of Economics and Political Science, 2014, http://eprints.lse.ac.uk/56706/1/Ending_the%20_drug_wars.pdf.

29 Sarah Sullivan, 'Support for Grass Grows: 4 Steps to Keeping Workplaces Safe With New Marijuana Laws', Lockton Companies, 2017, https://www.lockton.com/whitepapers/Sullivan_Legalizing_Marijuana_April_2017_lo_res.pdf.

31 'Like'

1 https://dharmacomics.com/about/.

2 Julian Morgans, 'The Inventor of the "Like" Button Wants You to Stop Worrying About Likes', Vice, 6 July 2017, https://www.vice.com/en_uk/article/mbag3a/the-inventor-of-the-like-button-wants-you-to-stop-worrying-about-likes.

3 Trevor Haynes blog, 'Dopamine, Smartphones & You: A battle for your time', 1 May 2018, http://sitn.hms.harvard.edu/flash/2018/dopamine-smartphones-battle-time/; Bethany Brookshire, 'Dopamine Is _____: Is it love? Gambling? Reward? Addiction?', 3 July 2013, http://www.slate.com/articles/health_and_science/science/2013/07/what_is_dopamine_love_lust_sex_addiction_gambling_motivation_reward.html; Adam Alter, *Irresistible* (New York: Penguin Books, 2017).

4 Morgans, 'The Inventor'.

5   Victor Luckerson, 'The Rise of the Like Economy', The Ringer, 15 February 2017, https://www.theringer.com/2017/2/15/16038024/how-the-like-button-took-over-the-internet-ebe778be2459.

6   https://www.quora.com/Whats-the-history-of-the-Awesome-Button-that-eventually-became-the-Like-button-on-Facebook.

7   Gayle Cotton, 'Gestures to Avoid in Cross-Cultural Business: In Other Words, "Keep Your Fingers to Yourself!"', *Huffington Post*, 13 June 2013, https://www.huffingtonpost.com/gayle-cotton/cross-cultural-gestures_b_3437653.html.

8   Morgans, 'The Inventor'.

9   Hannes Grassegger and Mikael Krogerus, 'The Data That Turned the World Upside Down', Motherboard, 28 January 2017, https://motherboard.vice.com/en_us/article/mg9vvn/how-our-likes-helped-trump-win.

10 Grassegger and Krogerus, 'The Data'.

11 Jacob Kastrenakes, 'Facebook will limit developers' access to account data', The Verge, 21 March 2018, https://www.theverge.com/2018/3/21/17148726/facebook-developer-data-crackdown-cambridge-analytica.

12 David Nield, 'You Probably Don't Know All the Ways Facebook Tracks You', 8 June 2017, https://fieldguide.gizmodo.com/all-the-ways-facebook-tracks-you-that-you-might-not-kno-1795604150.

13 Rob Goldman, 'Hard Questions: What Information Do Facebook Advertisers Know About Me?', Facebook, 23 April 2018, https://newsroom.fb.com/news/2018/04/data-and-advertising/.

14 Julia Angwin, Ariana Tobin and Madeleine Varner, 'Facebook (Still) Letting Housing Advertisers Exclude Users by Race', ProPublica, 21 November 2017, https://www.propublica.org/article/facebook-advertising-discrimination-housing-race-sex-national-origin.

15 Julia Angwin, Madeleine Varner and Ariana Tobin, 'Facebook Enabled Advertisers to Reach "Jew Haters"', ProPublica, 14 September 2017, https://www.propublica.org/article/facebook-enabled-advertisers-to-reach-jew-haters.

16 BBC, 'Facebook data: How it was used by Cambridge Analytica', BBC, https://www.bbc.co.uk/news/av/technology-43674480/facebook-data-how-it-was-used-by-cambridge-analytica.

17 Grassegger and Krogerus, 'The Data'.

18 Sam Machkovech, 'Report: Facebook helped advertisers target teens who feel "worthless"', Ars Technica. 1 May 2017, https://arstechnica.com/information-technology/2017/05/facebook-helped-advertisers-target-teens-who-feel-worthless/.

19 'Comments on Research and Ad Targeting', Facebook, 30 April 2017, https://newsroom.fb.com/news/h/comments-on-research-and-ad-targeting/.

20 'Facebook admits failings over emotion manipulation study', BBC, 3 October 2014, https://www.bbc.co.uk/news/technology-29475019.

21 Olivia Goldhill, 'The psychology behind Cambridge Analytica is massively

overhyped', Quartz, 29 March 2018, https://qz.com/1240331/cambridge-analytica-psychology-the-science-isnt-that-good-at-manipulation/.

22 Mark Irvine, 'Facebook Ad Benchmarks for YOUR Industry [Data]', The Wordstream Blog, 28 February 2017, https://www.wordstream.com/blog/ws/2017/02/28/facebook-advertising-benchmarks.

32 Cassava Processing

1 https://www.damninteresting.com/the-curse-of-konzo/; Geoff Watts, 'Hans Rosling: Obituary', Lancet, 389.18 (February 2017), https://www.thelancet.com/pdfs/journals/lancet/PIIS0140-6736(17)30392-6.pdf.

2 J. Henrich and R. McElreath, 'The evolution of cultural evolution', Evolutionary Anthropology: Issues, News, and Reviews 12.3 (2003): 123–35, https://henrich.fas.harvard.edu/files/henrich/files/henrich_mcelreath_2003.pdf.

3 http://www.burkeandwills.net.au/Brief_History/Chapter_15.htm.

4 Jared Diamond, Guns, Germs and Steel (New York: WW Norton, 2005), p. 296.

5 Joseph Henrich, The Secret of Our Success (Woodstock: Princeton University Press, 2016), ch. 3. See also Robert Boyd and Peter J. Richerson, The Origin and Evolution of Cultures (New York: Oxford University Press, 2005).

6 Cornell College of Agriculture and Life Sciences, https://poisonousplants.ansci.cornell.edu/toxicagents/thiaminase.html.

7 http://www.abc.net.au/science/articles/2007/03/08/2041341.htm; Henrich, The Secret, ch. 3.

8 http://www.fao.org/docrep/009/x4007e/X4007E04.htm#ch3.2.1.

9 Peter Longerich, Holocaust: The Nazi Persecution and Murder of the Jews (New York: Oxford University Press, 2010), pp. 281–2.

10 Henrich, The Secret, ch. 7.

11 Hipólito Nzwalo and Julie Cliff, 'Konzo: From Poverty, Cassava, and Cyanogen Intake to Toxico-Nutritional Neurological Disease', PLOS Neglected Tropical Diseases, June 2011. https://www.ncbi.nlm.nih.gov/pmc/articles/PMC3125150/.

12 Amy Maxmen, 'Poverty plus a poisonous plant blamed for paralysis in rural Africa', https://www.npr.org/sections/thesalt/2017/02/23/515819034/poverty-plus-a-poisonous-plant-blamed-for-paralysis-in-rural-africa.

13 A. P. Cardoso, E. Mirione, M. Ernesto, F. Massaza, J. Cliff, M. R. Haque, J. H. Bradbury, 'Processing of cassava roots to remove cyanogens', Journal of Food Composition Analysis 18 (2005): 451–60.

14 Henrich, The Secret, ch. 7.

15 Henrich, The Secret. The quote is from p. 99.

16 Maxime Derex, Jean-François Bonnefon, Robert Boyd, Alex Mesoudi, 'Causal understanding is not necessary for the improvement of culturally evolving technology', https://psyarxiv.com/nm5sh/.

17 Henrich, The Secret, ch. 2.

18 For example, A. Whiten et al., 'Social Learning in the Real-World', PLOS ONE 11.7 (2016), https://doi.org/10.1371/journal.pone.0159920.

19 Tyler Cowen and Joseph Henrich in conversation, https://medium.com/conversations-with-tyler/joe-henrich-culture-evolution-weird-psychology-social-norms-9756a97850ce.

33 PENSIONS

1   Kim Hill and A.Magdalena Hurtado, *Aché Life History: The Ecology and Demography of a Foraging People* (London: Taylor & Francis, 1996), pp. 235–6.
2   Jared Diamond,*The World Until Yesterday: What Can We Learn from Traditional Societies?* (Harmondsworth: Penguin Books, 2012), pp. 215–16.
3   Diamond, *World Until Yesterday*, pp. 210, 227–8.
4   Diamond, *World Until Yesterday*, p. 234.
5   Robert L. Clark, Lee A. Craig and Jack W. Wilson, *A History of Public Sector Pensions in the United States* (Philadelphia: University of Pennsylvania Press, 2003).
6   Sarah Laskow, 'How Retirement Was Invented', *Atlantic*, 24 October 2014, https://www.theatlantic.com/business/archive/2014/10/how-retirement-was-invented/381802/.
7   *Social protection for older persons: Policy trends and statistics 2017–19*, International Labour Office, Social Protection Department, Geneva, 2018, available at https://www.ilo.org/wcmsp5/groups/public/---ed_protect/---soc_sec/documents/publication/wcms_645692.pdf.
8   World Bank, *Averting the Old Age Crisis: Policies to Protect the Old and Promote Growth* (1994) describes itself as 'the first comprehensive, global examination of this complex and pressing set of issues'.
9   OECD, *Pensions at a Glance, 2011* (2011), Figure 1.3, available at https://www.oecd-ilibrary.org/docserver/pension_glance-2011-5-en.pdf.
10 https://www.oecd-ilibrary.org/economics/oecd-factbook-2015-2016/total-fertility-rates_factbook-2015-table3-en.
11 World Economic Forum, 'We'll Live to 100 – How Can We Afford It?', May 2017, available at: http://www3.weforum.org/docs/WEF_White_Paper_We_Will_Live_to_100.pdf.
12 *Economist, Falling Short. Pensions Special Report*, 9 April 2011, p. 7.
13 *Economist, Falling Short*, p. 1.
14 OECD, *Financial Incentives and Retirement Savings* (2018), available at: https://doi.org/10.1787/9789264306929-en.
15 'We'll Live to 100'.
16 https://www.youtube.com/watch?v=mS9LCR5P5wI.
17 Henrik Cronqvist, Richard H. Thaler and Frank Yu, *When Nudges Are Forever: Inertia in the Swedish Premium Pension Plan,* AEA Papers and Proceedings 108 (May 2018).
18 See e.g. World Economic Forum, *Investing in (and for) Our Future* (2019), p. 21, available at: http://www3.weforum.org/docs/WEF_Investing_in_our_Future_report_2019.pdf.
19 Diamond, *World Until Yesterday*, p. 214.

34 QWERTY

1 Koichi Yasuoka and Motoko Yasuoka, 'On the Prehistory of QWERTY', *ZINBUN* 42 (2011): 161–74, https://doi.org/10.14989/139379.

2 Paul David, 'Clio and the Economics of QWERTY', *American Economic Review* 75 (May 1985): 332–7.

3 Jimmy Stamp, 'Fact of fiction? The Legend of the QWERTY keyboard', *Smithsonian Magazine* 3 (May 2013), https://www.smithsonianmag.com/arts-culture/fact-of-fiction-the-legend-of-the-qwerty-keyboard-49863249/.

4 Stan Liebowitz and Stephen Margolis, 'The Fable of the Keys', *Journal of Law & Economics* XXXIII (April 1990), https://www.utdallas.edu/~liebowit/keys1.html.

5 Victor Keegan, 'Will MySpace Ever Lose Its Monopoly?', *Guardian*, 8 February 2007, https://www.theguardian.com/technology/2007/feb/08/business.comment.

35 The Langstroth Hive

1 Bernard Mandeville, *The Fable of the Bees or Private Vices, Publick Benefits*, vol. 1 (1732).

2 James Meade, 'External Economics and Diseconomies in a Competitive Situation', *Economic Journal* 62.245 (1952), https://www.jstor.org/stable/2227173.

3 Bee Wilson, *The Hive: The Story of the Honeybee and Us* (London: John Murray, 2004).

4 Randal Rucker and Walter Thurman, 'Colony Collapse Disorder: The Market Response to Bee Disease', *PERC Policy Series* 50 (2012).

5 https://patents.google.com/patent/US9300A/en.

6 Wilson, *Hive,* pp. 222–5.

7 Steven N. S. Cheung, 'The Fable of the Bees: An Economic Investigation', *Journal of Law and Economics* 16.1 (1973): 11–33.

8 *Economic Impacts of the California Almond Industry* (University of California Agricultural Issues Center), Appendix 2, http://aic.ucdavis.edu/almonds/Economic%20Impacts%20of%20California%20Almond%20Industry_Full%20Report_FinalPDF_v2.pdf .

9 Byard Duncan, 'California's almond harvest has created a golden opportunity for bee thieves', *Reveal News*, 8 October 2018, https://www.revealnews.org/article/californias-almond-harvest-has-created-a-golden-opportunity-for-bee-thieves/.

10 Sources vary as to how many. An article in *Scientific American* says between 20 and 80 billion, depending on various assumptions, https://www.scientificamerican.com/article/migratory-beekeeping-mind-boggling-math/; Dave Goulson's *Bee Quest* (London: Jonathan Cape, 2017) puts the number at 80 billion. Professor Goulson is a bumblebee expert but gives no source for this number.

11 Wilson, *Hive*, p. 54.

12 Goulson, *Bee Quest*, pp. 115–20.
13 Shawn Regan, 'How Capitalism Saved The Bees', https://www.perc. org/2017/07/20/how-capitalism-saved-the-bees/; Econtalk podcast: Wally Thurman on bees, beekeeping and Coase, http://www.econtalk.org/wally-thurman-on-bees-beekeeping-and-coase/ 16 Dec 2013. Also the House of Commons Library report on the UK Bee Population, published 10 November 2017, concluded that wild bee populations in the UK were in decline, while managed populations of honeybees and bumblebees were expanding, https://researchbriefings.parliament.uk/ResearchBriefing/ Summary/CDP-2017-0226.
14 https://www.gov.uk/government/news/a-boost-for-bees-900-million-countryside-stewardship-scheme.

36 Dams

1 Norman Smith, *A History of Dams* (London: Peter Davies, 1971), http:// www.hydriaproject.info/en/egypt-sadd-al-kafara-dam/waterworks22/.
2 'The Ups and Downs of Dams', *Economist*, 22 May 2010, https://www. economist.com/special-report/2010/05/22/the-ups-and-downs-of-dams.
3 *BP Statistical Review of World Energy* (2019), p. 9, https://www. bp.com/content/dam/bp/business-sites/en/global/corporate/pdfs/ energy-economics/statistical-review/bp-stats-review-2019-full-report.pdf.
4 http://en.people.cn/200510/01/eng20051001_211892.html.
5 Philip Ball, *The Water Kingdom* (London: Vintage, 2016), pp. 226–9.
6 Smith, *Dams*.
7 Charles Perrow, *Normal Accidents* (Chichester: Princeton University Press, 1999); Matthys Levy and Mario Salvadori, *Why Buildings Fall Down* (New York: WW Norton, 2002).
8 Protocol Additional to the Geneva Conventions of 12 August 1949, and relating to the Protection of Victims of International Armed Conflicts (Protocol I), 8 June 1977. Article 56, https://ihl-databases.icrc.org/ihl/ WebART/470-750071.
9 Benedict Mander, 'Brazil's Itaipú dam treaty with Paraguay up for renewal', *Financial Times*, 20 September 2017, https://www.ft.com/content/bf02af96-7eb8-11e7-ab01-a13271d1ee9c.
10 'Ups and Downs', *Economist*.
11 120,000, according to both anthropologist Thayer Scudder – https://link. springer.com/book/10.1007%2F978-981-10-1935-7 – and Smith, *Dams*. The National Geographic Society puts the figure much lower, at 50,000, https://www.nationalgeographic.org/thisday/jul21/aswan-dam-completed/.
12 Elinor Ostrom, 'Incentives, Rules of the Game, and Development', Annual Bank Conference of Development Economics, World Bank, May 1995.
13 Esther Duflo and Rohini Pande, 'Dams', *Quarterly Journal of Economics*, MIT Press 122.2 (2007): 601–46.
14 Sheila M. Olmstead and Hilary Sigman, 'Damming the Commons: An Empirical Analysis of International Cooperation and Conflict in Dam

Location', *Journal of the Association of Environmental and Resource Economists*, University of Chicago Press 2.4 (2015): 497–526.

15 Heba Saleh and Tom Wilson, 'Tensions rise between Ethiopia and Egypt over use of river Nile', *Financial Times*, 20 October 2019, https://www.ft.com/content/b0ae7a52-f18d-11e9-ad1e-4367d8281195.

16 Asit K. Biswas, 'Aswan Dam Revisited: The Benefits of a Much-Maligned Dam', *Development and Cooperation* 6 (November/December 2002): 25–7, https://www.icid.org/aswan_paper.pdf; and 'The Aswan High Dam', https://www.water-technology.net/projects/aswan-high-dam-nile-sudan-egypt/.

17 Duflo and Pande, 'Dams'.

37 Fire

1 E. C. Pulaski, 'Surrounded by Forest Fires: My Most Exciting Experience as a Forest Ranger', American Forestry, available at https://foresthistory.org/wp-content/uploads/2017/02/Surrounded-by-Forest-Firest-By-E.C.-Pulaski.pdf.

2 Pulaski, 'Surrounded'.

3 The Great Fire of 1910. Available at: https://www.fs.usda.gov/Internet/FSE_DOCUMENTS/stelprdb5444731.pdf.

4 Pulaski, 'Surrounded'.

5 Andrew C. Scott, David M. J. S. Bowman, William J. Bond, Stephen J. Pyne, Martin E. Alexander, *Fire on Earth – an Introduction* (Chichester: Wiley-Blackwell, 2014).

6 Andrew C. Scott, *Burning Planet: The Story of Fire Through Time* (Oxford: Oxford University Press, 2018).

7 Charles Q. Choi, 'Savanna, Not Forest, Was Human Ancestors' Proving Ground', 3 August 2011, https://www.livescience.com/15377-savannas-human-ancestors-evolution.html.

8 'I Wan'na Be Like You (The Monkey Song)', lyrics available at: http://disney.wikia.com/wiki/I_Wan%27na_Be_Like_You.

9 Dennis Sandgathe and Harold L. Dibble, 'Who Started the First Fire?', 26 January 2017, https://www.sapiens.org/archaeology/neanderthal-fire/.

10 J. A. J. Gowlett, 'The discovery of fire by humans: a long and convoluted process', available at: http://rstb.royalsocietypublishing.org/content/371/1696/20150164.

11 Martha Carney, 'Local knowledge says these raptors hunt with fire', 25 February 2018, https://www.futurity.org/firehawks-fire-birds-1687992-2/.

12 Sandgathe and Dibble, 'Who Started'.

13 Gowlett, 'The discovery'.

14 Scott, *Burning Planet*.

15 Richard Wrangham, *Catching Fire: How Cooking Made Us Human* (London: Profile Books, 2009).

16 J. A. J. Gowlett, 'Firing Up the Social Brain', University of Liverpool,

Proceedings of the British Academy 158 (January 2012), https://www.researchgate.net/publication/281717936_Firing_Up_the_Social_Brain.

17 Stephen J. Pyne, 'The Fire Age', 5 May 2015, https://aeon.co/essays/how-humans-made-fire-and-fire-made-us-human.

18 World Health Organization, 'Household air pollution and health', 8 May 2018, http://www.who.int/news-room/fact-sheets/detail/household-air-pollution-and-health.

19 Scott, *Burning Planet.*

20 Scott, *Burning Planet.*

21 Greg Ip, *Foolproof: Why Safety Can Be Dangerous and How Danger Makes Us Safe* (London: Hachette UK, 2015).

38 Oil

1 Lisa Margonelli, *Oil on the Brain* (New York: Penguin Random House, 2007), p. 285.

2 'Pithole's Rise And Fall', *New York Times*, 26 December 1879, https://timesmachine.nytimes.com/timesmachine/1879/12/26/80704720.pdf.

3 Matthew Yeomans, *Oil* (New York: New Press, 2004), pp. xvi–xvii.

4 *BP Statistical Review of World Energy 2018*, p. 9, https://www.bp.com/content/dam/bp/business-sites/en/global/corporate/pdfs/energy-economics/statistical-review/bp-stats-review-2018-full-report.pdf.

5 Eliot Jones, *The Trust Problem in the United States* (New York: Macmillan, 1921), p. 47, https://archive.org/details/trustprobleminu00jonegoog/page/n72.

6 Maria Gallucci, 'Container Ships Use Super-Dirty Fuel. That Needs To Change', *Wired*, 9 November 2017, https://www.wired.com/story/container-ships-use-super-dirty-fuel-that-needs-to-change/.

7 James Hamilton, 'Oil Shocks and Recession', *Econbrowser*, April 2009, http://econbrowser.com/archives/2009/04/oil_shocks_and_1; and Justin Lahart, 'Did The Oil Price Boom Of 2008 Cause Crisis?', *Wall Street Journal*, 3 April 2009, https://blogs.wsj.com/economics/2009/04/03/did-the-oil-price-boom-of-2008-cause-crisis/.

8 Daniel Yergin, *The Prize* (London: Simon and Schuster, 1991), pp. 11–12.

9 'Britain Fights Oil Nationalism', *New York Times Archive*, https://archive.nytimes.com/www.nytimes.com/library/world/mideast/041600iran-cia-chapter1.html.

10 Javier Blas and Will Kennedy, 'Saudi Aramco's $2 Trillion Zombie IPO', *Bloomberg*, 7 July 2018, https://www.bloomberg.com/news/articles/2018-07-07/saudi-aramco-s-2-trillion-zombie-ipo.

11 Anthony J. Venables, 'Using Natural Resources for Development: Why Has It Proven So Difficult?', *Journal of Economic Perspectives* 30.1: 161–84, doi:10.1257/jep.30.1.161; and Michael Ross, 'What Have We Learned about the Resource Curse?', *Annual Review of Political Science* 18 (2015): 239–59, doi:10.1146/annurev-polisci-052213-040359.

12 Alexandra Starr, 'Caracas: Living Large On Oil', *American Scholar*, 1 March

2007, https://theamericanscholar.org/letter-from-caracas/#.XFg50lz7SUk.

13 Bill Gates, 'Beating Nature at Its Own Game', *Gates Notes*, 14 March 2018, https://www.gatesnotes.com/Energy/Beating-Nature.

14 Spencer Dale, 'New Economics of Oil', Speech to the Society of Business Economists Annual Conference, 13 October 2015.

39 Vulcanisation

1 Sharon Sliwinski, *The Kodak on the Congo* (London: Autograph ABP, 2010), available at: https://www.academia.edu/2464487/ In_the_early_1900s_the_missionaries_Alice_Seeley_Harris.

2 Adam Hochschild, *King Leopold's Ghost* (New York/Boston: Mariner Books, 1999), p. 120.

3 Sliwinski, *Kodak on the Congo*.

4 Hochschild, *Leopold's Ghost*, p. 120.

5 Bradford Kinney Peirce, *Trials of an Inventor: Life and Discoveries of Charles Goodyear* (New York: Carlton & Porter, 1868).

6 Charles Sack, *Noble Obsession: Charles Goodyear, Thomas Hancock, and the Race to Unlock the Greatest Industrial Secret of the Nineteenth Century* (New York: Hyperion, 2002).

7 Sack, *Noble Obsession*.

8 Sack, *Noble Obsession*.

9 Sack, *Noble Obsession*.

10 Hochschild, *Leopold's Ghost*, p. 158.

11 Sack, *Noble Obsession*.

12 Hochschild, *Leopold's Ghost*, p. 158.

13 Hochschild, *Leopold's Ghost*, pp. 160–2.

14 World Rubber Industry, 23 June 2016, https://www.prnewswire.com/ news-releases/world-rubber-industry-300289614.html.

15 U.S. Synthetic Rubber Program, American Chemical Society, https:// www.acs.org/content/acs/en/education/whatischemistry/landmarks/ syntheticrubber.html.

16 Sheldon Brown and John Allen, 'Bicycle Tires and Tubes', https://www. sheldonbrown.com/tires.html.

17 Michelle Labbe, 'Properties of Natural & Synthetic Rubber', Sciencing, https://sciencing.com/properties-natural-synthetic-rubber-7686133.html.

18 Charles C. Mann, 'Why We (Still) Can't Live Without Rubber', *National Geographic*, December 2015, https://www.nationalgeographic.com/ magazine/2016/01/southeast-asia-rubber-boom/.

19 Mbom Sixtus, 'Indigenous communities at risk as Chinese rubber firm uses land', 10 December 2018, https://www.aljazeera.com/indepth/ features/indigenous-communities-risk-chinese-rubber-firm-land- 181209211730629.html.

40 The Wardian Case

1   Robert Fortune, *Three Years Wanderings in the Northern Provinces of China* (London: Spottiswoode and Shaw, 1847), available at: http://www.gutenberg.org/files/54720/54720-h/54720-h.htm.

2   Nathaniel Bagshaw Ward, *On the Growth of Plants in Closely Glazed Cases* (London: John van Voorst, 1842).

3   Maggie Campbell-Culver, *The Origin of Plants: The people and plants that have shaped Britain's garden history since the year 1000* (London: Transworld, 2001).

4   Ward, *Growth of Plants*.

5   Ward, *Growth of Plants*.

6   Toby Musgrave, Chris Gardner, Will Musgrave, *The Plant Hunters. Two Hundred Years of Adventure and Discovery Around the World* (London: Seven Dials, 1999).

7   Musgrave et al. *The Plant Hunters*.

8   Christopher Thacker, *The History of Gardens* (Berkeley, CA: University of California Press, 1979).

9   https://en.wikipedia.org/wiki/William_Cavendish,_6th_Duke_of_Devonshire.

10 R. R. Resor, 'Rubber in Brazil: Dominance and Collapse, 1876–1945', *Business History Review* 51.03 (1977): 341–66, doi:10.2307/3113637.

11 Sarah Rose, *For All the Tea in China: Espionage, Empire and the Secret Formula for the World's Favourite Drink* (London: Random House, 2013).

12 Rose, *For All the Tea*.

13 Fortune, *Three Years Wanderings*.

14 Luke Keogh, 'The Wardian Case: How a Simple Box Moved the Plant Kingdom', *Arnoldia* 74.4 (May 2017), Arnold Arboretum of Harvard University, available at: http://arnoldia.arboretum.harvard.edu/pdf/issues/2017-74-4-Arnoldia.pdf.

15 Keogh, 'Wardian Case'.

16 Daniel R. Headrick, *The Tools of Empire: Technology and European Imperialism in the Nineteenth Century* (Oxford: Oxford University Press, 1981).

41 Cellophane

1   'You're The Top', Cole Porter, lyrics at: https://www.lyricsmode.com/lyrics/c/cole_porter/youre_the_top.html.

2   'Plastic-wrapped bananas and the "kiwi spoon": your packaging peeves', *Guardian*, 29 August 2017, https://www.theguardian.com/sustainable-business/2017/aug/29/plastic-packaging-peeves-straws-avocados-single-use-waste-supermarkets-your-photos.

3   'Are seafood lovers really eating 11,000 bits of plastic per year?', Reality Check team, BBC News, 17 December 2017, https://www.bbc.co.uk/news/science-environment-42270729.

4   Stephen Fenichell, *Plastic: The Making of a Synthetic Century* (London: HarperBusiness, 1996).

5   Heather S. Morrison, *Inventors of Food and Agriculture Technology* (New York: Cavendish Square Publishing, 2015).

6   Ai Hisano, *Cellophane, the New Visuality, and the Creation of Self-Service Food Retailing*, Harvard Business School Working Paper (2017): 17–106.

7   David A. Hounshell, John Kenly Smith, Jr, Victor Smith, *Science and Corporate Strategy: Du Pont R and D, 1902–1980* (Cambridge: Cambridge University Press, 1988).

8   *Inventors and Inventions*, vol. 1, Marshall Cavendish, 2008.

9   Hounshell et al., *Science and Corporate Strategy*.

10 Hisano, *Cellophane*.

11 Craig Davidson and Fred Orval Briton, *How to Make Money Selling Meat* (The Progressive Grocer, 1937), cited in Hisano, *Cellophane*.

12 Hisano, *Cellophane*.

13 Mary Bellis, 'The Inventor of Saran Wrap', Thoughtco, 19 November 2019, https://www.thoughtco.com/history-of-pvdc-4070927.

14 Alan Greene, *Raising Baby Green: The Earth-Friendly Guide to Pregnancy, Childbirth, and Baby Care* (San Francisco: John Wiley & Sons, 23 Dec 2010), p. 151, https://books.google.co.uk/books?id=GstzPDifvsIC&pg=PA151.

15 https://en.wikipedia.org/wiki/Phase-out_of_lightweight_plastic_bags.

16 'Types of Plastic Packaging', The Waste and Resources Action Programme (WRAP), http://www.wrap.org.uk/collections-and-reprocessing/dry-materials/plastics/guidance/types-plastic-packaging.

17 Alexander H. Tullo, 'The cost of plastic packaging', *Chemical & Engineering News* 94.41 (17 October 2016): 32–7, https://cen.acs.org/articles/94/i41/cost-plastic-packaging.html.

18 See e.g. Lars G. Wallentin, 'Multi-layer materials', 6 January 2018, http://www.packagingsense.com/2018/01/06/multi-layer-materials/ and Tom Szaky, 'Is less packaging really good for the environment?', 12 March 2015, https://www.weforum.org/agenda/2015/03/is-less-packaging-really-good-for-the-environment/.

19 *Packaging in Perspective*, Advisory Committee on Packaging, October 2008, http://webarchive.nationalarchives.gov.uk/20130403095620/http://archive.defra.gov.uk/environment/waste/producer/packaging/documents/packaginginperspective.pdf.

20 *Packaging in Perspective*.

21 Sam Knight, 'Plastic – The Elephant In The Room', *Financial Times*, 26 April 2008.

22 Knight, 'Plastic'.

23 *Packaging in Perspective*.

24 *Life Cycle Assessment of grocery carrier bags*, Environmental Project no. 1985, The Danish Environmental Protection Agency (February 2018), https://www2.mst.dk/Udgiv/publications/2018/02/978-87-93614-73-4.pdf.

25 *Life Cycle Assessment*.

26 See e.g. *The New Plastics Economy: Rethinking the future of plastics*, Ellen MacArthur Foundation, 19 January 2016, https://www.

ellenmacarthurfoundation.org/publications/the-new-plastics-economy-rethinking-the-future-of-plastics.

42 Recycling

1   Dongguan Base, Nine Dragons Paper, http://www.ndpaper.com/en/business/dongguanbase.php.

2   Monica Sanders, 'Zhang Yin: The World's Richest Woman', https://www.legalzoom.com/articles/zhang-yin-the-worlds-richest-woman.

3   Evan Osnos, 'Wastepaper Queen', *New Yorker*, 23 March 2009, https://www.newyorker.com/magazine/2009/03/30/wastepaper-queen.

4   Adam Minter, 'Nine Dragons and a Whole Lotta Labor', 12 March 2008, http://shanghaiscrap.com/2008/03/nine-dragons-and-a-whole-lotta-labor/.

5   'A Chinese ban on rubbish imports is shaking up the global junk trade', *Economist*, 17 September 2018, https://www.economist.com/special-report/2018/09/29/a-chinese-ban-on-rubbish-imports-is-shaking-up-the-global-junk-trade.

6   'A Chinese ban'.

7   Bob Tita, 'Recycling, Once Embraced by Businesses and Environmentalists, Now Under Siege', *Wall Street Journal*, 13 May 2018, https://www.wsj.com/articles/recycling-once-embraced-by-businesses-and-environmentalists-now-under-siege-1526209200.

8   Leslie Hook and John Reed, 'Why the world's recycling system stopped working', *Financial Times*, 25 October 2018, https://www.ft.com/content/360e2524-d71a-11e8-a854-33d6f82e62f8.

9   https://en.wikipedia.org/wiki/Palimpsest.

10   Martin Medina, *The World's Scavengers: Salvaging for Sustainable Consumption and Production* (Lanham, MD): AltaMira Press, 2007), pp. 20–21.

11   Dard Hunter, *Papermaking: The History and Technique of an Ancient Craft*, 2nd ed (New York: Knopf, 1957), p. 54, quoted on 'Some of the Earliest Paper Recycling Occurred in Japan', Jeremy Norman's online History of Information, http://www.historyofinformation.com/expanded.php?id=3977.

12   Medina, *The World's Scavengers*, 2007), p. 70.

13   *Life*, 1 August 1955, available at: https://books.google.co.uk/books?id=xlYEAAAAMBAJ&pg=PA43.

14   Olivia B. Waxman, 'The History of Recycling in America Is More Complicated Than You May Think', *Time*, 15 November 2016, http://time.com/4568234/history-origins-recycling/.

15   https://www.youtube.com/watch?v=j7OHG7tHrNM.

16   Ginger Strand, 'The Crying Indian', *Orion* Magazine, 20 November 2008, https://orionmagazine.org/article/the-crying-indian/.

17   Finis Dunaway, 'The "Crying Indian" ad that fooled the environmental movement', *Chicago Tribune*, 21 November 2017, https://www.chicagotribune.com/news/opinion/commentary/ct-perspec-indian-crying-environment-ads-pollution-1123-20171113-story.html.

18   Michele Nestor, 'Facing the Reality of Recycling Economics', Waste360, 4
      August 2016, https://www.waste360.com/business/facing-reality-recycling-
      economics.

19   Dunaway, '"Crying Indian" ad'.

20   Monic Sun and Remi Trudel, 'The Effect of Recycling versus Trashing on
      Consumption: Theory and Experimental Evidence', Boston University,
      16 May 2016. Available at: https://www.researchgate.net/profile/
      Monic_Sun/publication/303263301_The_Effect_of_Recycling_versus_
      Trashing_on_Consumption_Theory_and_Experimental_Evidence/
      links/573a555d08ae9f741b2ca8e1/The-Effect-of-Recycling-versus-
      Trashing-on-Consumption-Theory-and-Experimental-Evidence.pdf.

21   John Tierney, 'The Reign of Recycling', New York Times, 3 October
      2015, https://www.nytimes.com/2015/10/04/opinion/sunday/
      the-reign-of-recycling.html.

22   Tierney, 'Reign of Recycling'.

23   Tita, 'Recycling'.

24   'Emerging economies are rapidly adding to the global pile of garbage',
      Economist, 27 September 2018, https://www.economist.com/
      special-report/2018/09/29/emerging-economies-are-rapidly-adding-
      to-the-global-pile-of-garbage.

25   Towards the Circular Economy, World Economic Forum, January 2014,
      http://www3.weforum.org/docs/WEF_ENV_TowardsCircularEconomy_
      Report_2014.pdf.

26   https://www.bbc.co.uk/programmes/m0000t55.

27   Monica Nickelsburg, 'Meet the TrashBot: CleanRobotics is using machine
      learning to keep recycling from going to waste', 5 February 2018, https://
      www.geekwire.com/2018/meet-trashbot-cleanrobotics-using-machine-
      learning-keep-recycling-going-waste/.

28   Hook and Reed, 'World's recycling system'.

43 Dwarf Wheat

1   Noel Vietmeyer, Our Daily Bread: The Essential Norman Borlaug (Lorton:
     Bracing Books, 2011).

2   Dr Paul R. Ehrlich, The Population Bomb (New York: Ballantine
     Books, 1968).

3   Vietmeyer, Daily Bread.

4   Charles C. Mann, The Wizard and the Prophet: Two Groundbreaking Scientists and
     Their Conflicting Visions of the Future of Our Planet (New York: Picador, 2018).

5   http://www.worldometers.info/world-population/
     world-population-by-year/.

6   'Thomas Malthus (1766–1834)', BBC History, http://www.bbc.co.uk/
     history/historic_figures/malthus_thomas.shtml.

7   Thomas Malthus, An Essay on the Principle of Population, as it Affects the Future
     Improvement of Society with Remarks on the Speculations of Mr. Godwin, M.
     Condorcet, and Other Writers (London, 1798).

8   *World Population Prospects: The 2017 Revision*, 21 June 2017, UN Department
     of Economic and Social Affairs, https://www.un.org/development/desa/
     publications/world-population-prospects-the-2017-revision.html.

9   Mann, *The Wizard*.

10   Paul Ehrlich, 'Collapse of civilisation is a near certainty within decades',
     interview with Damian Carrington, *Guardian*, 22 March 2018, https://
     www.theguardian.com/cities/2018/mar/22/collapse-civilisation-near-certa
     in-decades-population-bomb-paul-ehrlich.

11   Mann, *The Wizard*.

12   Daniel Norero, 'GMO crops have been increasing yield for 20 years, with
     more progress ahead', Cornell Alliance for Science, 23 February 2018,
     https://allianceforscience.cornell.edu/blog/2018/02/gmo-crops-increasing-
     yield-20-years-progress-ahead/.

13   Eric Niiler, 'Why Gene Editing Is the Next Food Revolution', *National
     Geographic*, 10 August 2018, https://www.nationalgeographic.com/
     environment/future-of-food/food-technology-gene-editing/.

14   Mann, *The Wizard*.

44 Solar Photovoltaics

1   Ken Butti and John Perlin, *A Golden Thread: 2500 Years of Solar Architecture
     and Technology* (London: Marion Boyars, 1981).

2   Varun Sivaram, *Taming the Sun* (Cambridge, MA: MIT Press, 2018), p. 29.

3   Werner Weiss and Franz Mauthner, *Solar Heat Worldwide: Markets and
     Contribution to the Energy Supply 2010* (Gleisdorf: Institute for Sustainable
     Technologies, 2012), cited by Sivaram, *Taming*, p. 30.

4   Jon Gertner, *The Idea Factory* (London: Penguin, 2012), pp. 170–2.

5   Chris Goodall, *The Switch* (London: Profile, 2016).

6   Goodall, *The Switch*; also the International Renewable Energy Agency,
     http://www.irena.org/-/media/Images/IRENA/Costs/Chart/
     Solar-photovoltaic/fig-62.png.

7   T. P. Wright, 'Factors Affecting the Cost of Airplanes', *Journal of the
     Aeronautical Sciences* 3 (February 1936).

8   BCG Research summarised by Goodall in *The Switch*; also Martin
     Reeves, George Stalk and Filippo S. Pasini, 'BCG Classics Revisited: The
     Experience Curve', https://www.bcg.com/publications/2013/growth-
     business-unit-strategy-experience-curve-bcg-classics-revisited.aspx.

9   François Lafond, Aimee G. Bailey, Jan D. Bakker, Dylan Rebois, Rubina
     Zadourian, Patrick McSharry, J. Doyne Farmer, 'How well do experience
     curves predict technological progress?', *Technological Forecasting and Social
     Change*, 128 (2017), https://arxiv.org/abs/1703.05979.

10   Goodall, *The Switch*.

11   Sivaram, *Taming*, pp. 13–14.

12   Goodall, *The Switch*; Ed Crooks, 'US China solar duties fail to halt imports
     as EU prepares its move', *Financial Times*, 2 June 2013, https://www.ft.com/
     content/a97482e8-c941-11e2-bb56-00144feab7de.

13 Pilita Clark, 'The Big Green Bang', *Financial Times*, 18 May 2017, https://www.ft.com/content/44ed7e90-3960-11e7-ac89-b01cc67cfeec.

45 The Hollerith Punch-Card Machine

1 https://en.wikipedia.org/wiki/List_of_public_corporations_by_market_capitalization (accessed 26 June 2019).

2 See e.g. 'The world's most valuable resource is no longer oil, but data', *Economist*, 6 May 2017, https://www.economist.com/leaders/2017/05/06/the-worlds-most-valuable-resource-is-no-longer-oil-but-data.

3 https://en.wikipedia.org/wiki/List_of_public_corporations_by_market_capitalization (accessed 26 June 2019) – see figures for first quarter 2011.

4 Bernard Marr, 'Here's Why Data Is Not The New Oil', *Forbes*, 5 March 2018, https://www.forbes.com/sites/bernardmarr/2018/03/05/heres-why-data-is-not-the-new-oil/.

5 A typical video ad on YouTube in 2019 cost between $0.10 and $0.30, according to Betsy McLeod, 'How much does it cost to advertise on YouTube in 2019?', Blue Corona, 27 February 2018, https://www.bluecorona.com/blog/how-much-does-it-cost-to-advertise-youtube.

6 Geoffrey D. Austrian, *Herman Hollerith: Forgotten Giant of Information Processing* (New York: Columbia University Press, 1982).

7 United States Census Bureau, 'Census in the Constitution', https://www.census.gov/programs-surveys/decennial-census/about/census-constitution.html.

8 James R. Beniger, *The Control Revolution: Technical and Economic Origins of the Information Society* (Cambridge, MA: Harvard University Press, 1986), p. 409.

9 Beniger, *Control Revolution*, p. 412.

10 Beniger, *Control Revolution*, p. 412.

11 Austrian, *Herman Hollerith*.

12 Robert L. Dorman, 'The Creation and Destruction of the 1890 Federal Census', *American Archivist* 71 (Fall/Winter 2008): 350–83.

13 Beniger, *Control Revolution*, p. 416.

14 Beniger, *Control Revolution*, p. 416.

15 Austrian, *Herman Hollerith*.

16 Adam Tooze, *Statistics and the German State 1900–1945: The Making of Modern Economic Knowledge* (Cambridge: Cambridge University Press, 2008).

17 Beniger, *Control Revolution*, p. 408.

18 Edwin Black, *IBM and the Holocaust: The Strategic Alliance between Nazi Germany and America's Most Powerful Corporation* (Washington DC, Dialog Press, 2001).

19 Beniger, *Control Revolution*, pp. 420–1.

20 https://en.wikipedia.org/wiki/List_of_public_corporations_by_market_capitalization (accessed 26 June 2019) – see figures for first quarter 2013.

46 The Gyroscope

1   Sean A. Kingsley, *The Sinking of the First Rate Victory (1744): A Disaster Waiting to Happen?* (London: Wreck Watch Int., 2015), http://victory1744. org/documents/OMEPapers45_000.pdf.

2   https://www.telegraph.co.uk/history/11411508/Tory-Lord-defends-the-treasure-hunt-for-HMS-Victory.html.

3   Sylvanus Urban, *The Gentleman's Magazine and Historical Chronicle*, London, vol. XXIV, for the year MDCCLIV, https://books.google.co.uk/ books?id=0js3AAAAYAAJ&pg=PA447.

4   Urban, *Gentleman's Magazine.*

5   Urban, *Gentleman's Magazine.*

6   https://blog.sciencemuseum.org.uk/john-smeaton-whirling-speculum/.

7   Urban, *Gentleman's Magazine.*

8   Silvio A. Bedini, *History Corner: The Artificial Horizon*, in *Professional Surveyor Magazine* Archives online, http://atlantic-cable.com/Article/Combe/ ArtificialHorizon/article.idc.html.

9   Ljiljana Veljović, 'History and Present of Gyroscope Models and Vector Rotators', *Scientific Technical Review* 60.3–4 (2010): 101–111, http://www.vti. mod.gov.rs/ntp/rad2010/34-10/12/12.pdf.

10 Mario N. Armenise, Caterina Ciminelli, Francesco Dell'Olio, Vittorio M. N. Passaro, *Advances in Gyroscope Technologies* (Berlin: Springer, 2010).

11 'mCube Redefines MEMS Sensor Innovation by Unveiling the World's Smallest 1x1mm Accelerometer', mCube, 27 October 2015, http://www. mcubemems.com/news-events/press-releases/mcube-mc3571-pr/.

12 'Light-powered gyroscope is world's smallest: Promises a powerful spin on navigation', Nanowerk, 2 April 2015, https://www.nanowerk.com/ nanotechnology-news/newsid=39634.php.

13 'Remote Piloted Aerial Vehicles: An Anthology', http://www.ctie.monash. edu.au/hargrave/rpav_home.html#Beginnings.

14 'The history of drones and quadcopters', Quadcopter Arena, https:// quadcopterarena.com/the-history-of-drones-and-quadcopters/.

15 Andrew J. Hawkins, 'Ehang's passenger-carrying drones look insanely impressive in first test flights', The Verge, 5 February 2018, https://www. theverge.com/2018/2/5/16974310/ehang-passenger-carrying-drone-first-test-flight.

16 Jiayang Fang, 'How e-commerce is transforming rural China', *New Yorker*, 16 July 2018, https://www.newyorker.com/magazine/2018/07/23/how-e-commerce-is-transforming-rural-china.

17 Nicole Kobie, 'Droning on: the challenges facing drone delivery', http:// www.alphr.com/the-future/1004520/droning-on-the-challenges-facing-drone-delivery.

18 Neil Hughes, 'Startup plots drone-delivered packages that could securely fly in your window', 18 October 2016, https://oneworldidentity. com/startup-plots-drone-delivered-packages-that-could-securely-fly-in-your-window/.

19 'Can Amazon's Drones Brave Winter Storms?', PYMNTS, 1 January 2016, https://www.pymnts.com/in-depth/2016/can-amazons-drones-brave-winter-storms/.

47 Spreadsheets

1 Steven Levy, 'A Spreadsheet Way of Knowledge', Medium, 24 October 2014, https://medium.com/backchannel/a-spreadsheet-way-of-knowledge-8de60af7146e; *Harper's*, November 1984; *Planet Money*, 'Spreadsheets!', Episode 606, February 2015, http://www.npr.org/sections/money/2015/02/25/389027988/episode-606-spreadsheets.

2 Dan Bricklin's personal website: http://www.bricklin.com/jobs96.htm.

3 Levy, 'A Spreadsheet Way of Knowledge'.

4 Peter Davis, 'The Executive Computer: Lotus 1-2-3 faces up to the upstarts', *New York Times*, 13 March 1988, https://www.nytimes.com/1988/03/13/business/the-executive-computer-lotus-1-2-3-faces-up-to-the-upstarts.html.

5 Daniel and Richard Susskind, *The Future of the Professions* (Oxford: Oxford University Press, 2015), esp. ch. 2.

6 Stephen G. Powell, Kenneth R. Baker and Barry Lawson, 'A critical review of the literature on spreadsheet errors', *Decision Support Systems* 46.1 (December 2008): 128–38, http://dx.doi.org/10.1016/j.dss.2008.06.001.

7 Ruth Alexander, 'Reinhart, Rogoff . . . and Herndon: The student who caught out the profs', BBC, 20 April 2013, https://www.bbc.co.uk/news/magazine-22223190.

8 Lisa Pollack, 'A Tempest in a Spreadsheet', *Financial Times* Alphaville, https://ftalphaville.ft.com/2013/01/17/1342082/a-tempest-in-a-spreadsheet/; Duncan Robinson, 'Finance Groups Lack Spreadsheet Controls', *Financial Times*, 18 March 2013, https://www.ft.com/content/60cea058-778b-11e2-9e6e-00144feabdc0#axzz2YaLVTi2m.

48 The Chatbot

1 Robert Epstein, 'From Russia, with Love: How I Got Fooled (and Somewhat Humiliated) by a Computer', *Scientific American Mind* 18.5 (October/November 2007): 16–17.

2 A. M. Turing, 'Computing Machinery and Intelligence', *Mind* 59 (1950): 433–60.

3 Brian Christian, *The Most Human Human* (New York: Doubleday, 2011).

4 Elizabeth Lopatto, 'The AI That Wasn't', Daily Beast, 10 June 2014, https://www.thedailybeast.com/the-ai-that-wasnt-why-eugene-goostman-didnt-pass-the-turing-test.

5 Brian Christian, 'The Samantha Test', *New Yorker*, 30 December 2013.

6 Kenneth M. Colby, James B. Watt and John P. Gilbert, 'A Computer Method for Psychotherapy: Preliminary Communication', *Journal of Nervous and Mental Diseases* 142.2 (1966): 148.

7   Erin Brodwin, 'I spent 2 weeks texting a bot about my anxiety', *Business Insider*, 30 January 2018, https://www.businessinsider.com/therapy-chatbot-depression-app-what-its-like-woebot-2018-1; Dillon Browne, Meredith Arthur and Miriam Slozberg, 'Do Mental Health Chatbots Work?', Healthline, 6 July 2018, https://www.healthline.com/health/mental-health/chatbots-reviews.

8   Chris Baraniuk, 'How Talking Machines Are Taking Call Centre Jobs', BBC News, 24 August 2018.

9   Alastair Sharp and Allison Martell, 'Infidelity website Ashley Madison facing FTC probe, CEO apologizes', Reuters, 5 July 2016, https://www.reuters.com/article/us-ashleymadison-cyber-idUSKCN0ZL09J.

10 Christian, *Most Human Human*.

11 John Markoff, 'Automated Pro-Trump Bots Overwhelmed Pro-Clinton Messages, Researchers Say', *New York Times*, 17 November 2016, https://www.nytimes.com/2016/11/18/technology/automated-pro-trump-bots-overwhelmed-pro-clinton-messages-researchers-say.html.

12 Adam Smith, *The Wealth of Nations* (1776), available at https://www.ibiblio.org/ml/libri/s/SmithA_WealthNations_p.pdf.

13 David Autor, 'Why Are There Still So Many Jobs? The History and Future of Workplace Automation', *Journal of Economic Perspectives* 29.3 (Summer 2015): 3–30.

14 F. G. Deters and M. R. Mehl, 'Does Posting Facebook Status Updates Increase or Decrease Loneliness? An Online Social Networking Experiment', *Social Psychological and Personality Science* 4.5 (2012), 10.1177/1948550612469233, doi:10.1177/1948550612469233.

49 The CubeSat

1   Robert Smith, 'What Happened When "Planet Money" Went On A Mission To Adopt A Spacecraft', NPR, 30 January 2018, https://www.npr.org/2018/01/30/581930126/what-happened-when-planet-money-went-on-a-mission-to-adopt-a-spacecraft.

2   Clive Cookson, 'Nano-satellites dominate space and spread spies in the skies', *Financial Times*, 11 July 2016, https://www.ft.com/content/33ca3cba-3c50-11e6-8716-a4a71e8140b0.

3   Quoted in Leonard David, 'Cubesats: Tiny Spacecraft, Huge Payoffs', Space.com, 8 September 2004, https://www.space.com/308-cubesats-tiny-spacecraft-huge-payoffs.html.

4   John Thornhill, 'A Space Revolution: do tiny satellites threaten our privacy?', *Financial Times Magazine*, 17–18 February 2018, https://www.ft.com/content/c7e00344-111a-11e8-940e-08320fc2a277; R. S. Jakhu and J. N. Pelton, 'Small Satellites and Their Regulation', SpringerBriefs in Space Development, doi: 10.1007/978-1-4614-9423-2_3, Springer New York, 2014.

5   Swapna Krishna, 'The Rise of Nanosatellites', *The Week*, 25 April 2018, http://theweek.com/articles/761349/rise-nanosatellites.

6   Interview with Adam Storeygard, 11 July 2018.

7   Rocket Labs have been reported as offering a $100,000 launch for a 1U CubeSat: Jamie Smyth, 'Small satellites and big data: a commercial space race hots up', *Financial Times*, 24 January 2018, https://www.ft.com/content/32d3f95e-f6c1-11e7-8715-e94187b3017e. The rocket brokers Spaceflight were quoting $295,000 for a 3U CubeSat in July 2018: http://spaceflight.com/schedule-pricing/.

8   Samantha Mathewson, 'India Launches Record-Breaking 104 Satellites on Single Rocket', Space.com, 15 February 2017, https://www.space.com/35709-india-rocket-launches-record-104-satellites.html.

9   Jon Porter, 'Amazon will launch thousands of satellites to provide internet around the world', The Verge, 4 April 2019, https://www.theverge.com/2019/4/4/18295310/amazon-project-kuiper-satellite-internet-low-earth-orbit-facebook-spacex-starlink.

10   Nanoracks company website: http://nanoracks.com/about-us/our-history/.

11   'Space 2: Wait, Why Are We Going To Space?', *Planet Money*, 1 December 2017, https://www.npr.org/templates/transcript/transcript.php?storyId=566713606.

12   Dave Donaldson and Adam Storeygard, 'The View from Above: Applications of Satellite Data in Economics', *Journal of Economic Perspectives* 30.4 (Fall 2016): 171–98, https://www.aeaweb.org/articles?id=10.1257/jep.30.4.171.

13   Interview with Josh Bumenstock, 10 July 2018.

50 The Slot Machine

1   Natasha Dow Schüll, *Addiction by Design: Machine Gambling in Las Vegas* (Woodstock: Princeton University Press, 2012).

2   Clifford Geertz, *The Interpretation of Cultures: Selected Essays* (New York: Basic Books, 1973).

3   Tim Harford, *The Logic of Life* (New York: Random House, 2008).

4   Rob Davies, 'Maximum stake for fixed-odds betting terminals cut to £2', *Guardian*, 17 May 2018, https://www.theguardian.com/uk-news/2018/may/17/maximum-stake-for-fixed-odds-betting-terminals-cut-to-2.

5   Schüll, *Addiction by Design*.

6   Alexander Smith, 'Historical Interlude: The History of Coin-Op Part 2, From Slot Machines to Sportslands', They Create Worlds blog, 25 March 2015, https://videogamehistorian.wordpress.com/2015/03/25/historical-interlude-the-history-of-coin-op-part-2-from-slot-machines-to-sportlands/.

7   'No-armed Bandit', *99 Percent Invisible*, Episode 78, 30 April 2013, https://99percentinvisible.org/episode/episode-78-no-armed-bandit/.

8   University of Waterloo Gambling Research Lab video, 'Losses Disguised as Wins', 22 January 2013, https://uwaterloo.ca/gambling-research-lab/about/video-stories.

9   C. Graydon, M. J. Dixon, M. Stange and J. A. Fugelsang, 'Gambling despite financial loss – the role of losses disguised as wins in multi-line slots', *Addiction* 114 (2019): 119–24, https://doi.org/10.1111/add.14406.

10   March Cooper, 'Sit and Spin: How slot machines give gamblers the business', *Atlantic*, December 2005, https://www.theatlantic.com/magazine/archive/2005/12/sit-and-spin/304392/.

11   Lauren Slater, *Opening Skinner's Box* (London: Bloomsbury, 2004).

12   R. B. Breen and M. Zimmerman, 'Rapid Onset of Pathological Gambling in Machine Gamblers'. *Journal of Gambling Studies* 18.1 (Spring 2002): 31–43, doi:10.1023/A:1014580112648.

13   Nathan Lawrence, 'The Troubling Psychology of Pay-to-Loot Systems', IGN, 24 April 2017, https://uk.ign.com/articles/2017/04/24/the-troubling-psychology-of-pay-to-loot-systems.

14   Jackson Lears, *Something for Nothing* (New York: Viking, 2003).

51 Chess Algorithms

1   'Kasparov vs Turing', University of Manchester press release, 26 June 2012, https://www.manchester.ac.uk/discover/news/kasparov-versus-turing/.

2   Frederic Friedel and Garry Kasparov, 'Reconstructing Turing's "Paper Machine"', ChessBase, 23 September 2017, https://en.chessbase.com/post/reconstructing-turing-s-paper-machine.

3   https://twobithistory.org/2018/08/18/ada-lovelace-note-g.html.

4   Donald E. Knuth, 'Ancient Babylonian Algorithms', *Communications of the ACM* 15.7 (July 1972): 671–7; Jeremy Norman, 'Ancient Babylonian Algorithms: The Earliest Programs', http://www.historyofinformation.com/detail.php?id=3920.

5   http://mathworld.wolfram.com/EuclideanAlgorithm.html.

6   Christopher Steiner, *Automate This* (New York: Portfolio Penguin, 2012); Claude E. Shannon, 'A Symbolic Analysis of Relay and Switching Circuits', *Transactions of the American Institute of Electrical Engineers* 57.12 (December 1938): 713–23.

7   Claude E. Shannon, 'Programming a Computer for Playing Chess', *Philosophical Magazine* series 7, 41.314 (March 1950).

8   Douglas Hofstadter, *Gödel, Escher, Bach: An Eternal Golden Braid* (New York: Basic Books, 1979).

9   https://video.newyorker.com/watch/chess-grandmaster-garry-kasparov-replays-his-four-most-memorable-games/ – from around 5 minutes in.

10   James Somers, 'The Man Who Would Teach Machines to Think', *Atlantic*, November 2013, https://www.theatlantic.com/magazine/archive/2013/11/the-man-who-would-teach-machines-to-think/309529/.

11   https://vqa.cloudcv.org/.

12   AI Index Report, 2019, https://hai.stanford.edu/ai-index/2019.

13   Hannah Kuchler, 'Google AI system beats doctors in detection tests for breast cancer', *Financial Times*, 1 January 2020, https://www.ft.com/content/3b64fa26-28e9-11ea-9a4f-963f0ec7e134; Daniel Susskind, *A World Without Work* (London: Allen Lane, 2020).

14   David H. Autor, Frank Levy and Richard J. Murnane, 'The skill content of recent technological change: An empirical exploration', *Quarterly Journal of Economics* 118.4 (2003): 1279–1333; Susskind, *World Without Work*.

15 Garry Kasparov, 'Chess, a *Drosophila* of reasoning', *Science*, 7
    December 2018.
16 James Somers, 'How the artificial intelligence program AlphaZero mastered
    its games', *New Yorker*, 28 December 2018, https://www.newyorker.com/
    science/elements/how-the-artificial-intelligence-program-alphazero-
    mastered-its-games.

About the Author

Tim Harford is a senior columnist for the *Financial Times*, and the presenter of *Cautionary Tales* and Radio 4's *More or Less*. He is an honorary fellow of the Royal Statistical Society, a member of Nuffield College, Oxford, and the winner of numerous awards for economic and statistical journalism. In 2019 he was made an OBE 'for services to improving economic understanding'. Tim lives in Oxford with his wife and three children. His other books include *The Undercover Economist*, *Adapt* and *Messy*.

Acknowledgements

In acknowledging the ideas that contributed to the first Fifty Things book, I was consumed with anxiety that I had forgotten some brilliant contribution. Nobody has yet complained about being left out, but I feel the same paranoia and, again, beg the forgiveness of anyone I have omitted.

I do recall with pleasure useful conversations and exchanges with David Bodanis, Mohamed El-Erian, Peter Eso, Edward Hadas, Mark Henstridge, Bohdanna Kesala, Paul Klemperer, James Kynge, Denise Lievesley, Helen Margetts, Charlotte McDonald, Katharina Rietzler, Martin Sandbu, Xa Sturgis, Jamie Walsh and Peyton Young. I am grateful to you all for your insights and your generosity in sharing them.

I am doubly grateful to the economists, historians, journalists and others who have provided so much of both the original reporting and the academic ideas that underpin this project. It is quite impossible to become an expert in the history and the consequences of fifty-one distinct inventions or ideas. My intellectual debt to others is evident, and a glance at the references should make it clear who those others are.

At Little Brown, Tim Whiting and particularly Nithya Rae coped superbly – again! – with tight deadlines and late copy. My agents Sue Ayton, Helen Purvis and Sally Holloway steered a potentially fraught project into a safe harbour with diplomacy and determination.

At the BBC, my producer Ben Crighton has been even more fun to work with than in the last series – deft, subtle and supportive. Thank you for doing your very best to make me sound good. Many others across the BBC have carried the project forward, including James Beard, Jennifer Clarke, Jon Manel, Janet Staples and my editor, Richard Vadon.

As always I'm grateful to my editors at the *Financial Times* – in particular Jonathan Derbyshire, Brooke Masters and Alec Russell – for their support and indulgence. The FT remains an inspiring home and I feel so fortunate to be allowed to be part of it.

But again, the most important collaborator on this project has been Andrew Wright, who researched and drafted many of the chapters with wisdom and humour, as well as vastly improving my own work. A brilliant colleague and an even better friend.

Thank you to my children, Stella, Africa and Herbie. You guys were no help at all, but you're awesome. And to Fran Monks: thank you for being on my team.

Index